Coal to Diamonds

true stories of triumph

Good Catch Publishing

Published in Beaverton, Oregon, by Good Catch Publishing. www.goodcatchpublishing.com

V1.1

Printed in the United States of America

Church Directory

Christian and Missionary Alliance Church
Rev. Matt Knighton
327 Madison Street
South Cle Elum, WA 98943
cma-church.com
509.674.2664

Cle Elum Community Church
Dr. Worth Wilson
201 E. 3rd Street
Cle Elum, WA 98922
cleelumcommunitychurch.org
509.674.2472

Mt. Pisgah Presbyterian Church
Dr. Worth Wilson
207 N. 1st Street
Roslyn, WA 98941
Roslynchurch.org
509.674.2472

Christian Life Assembly
Rev. Fred Boyd
514 E. 3rd Street
Cle Elum, WA 98922
email: ceclag@hotmail.com
509.674.5249

Table of contents

Acknowledgements

A good idea—no, a really *great* one—is often followed by a negative thought: "That's fantastic! But there's no way we can pull it off. We don't have enough people, or people with the right skills. It costs way too much. We don't know how…."

Daren Lindley, Director of Good Catch Publishing, convinced three local pastors it could be done: write a shared book of inspiring stories. An opportunity was born, and we took a step of faith! Gunnar Falk, our publishing consultant, brought fresh ideas and encouragement every step of the way. These men, and their entire staff, were conduits of possibility for this book.

One colorful highlight of this project was that these three very different men, from very different backgrounds, who pastor four very different congregations, were able to work together so agreeably. They share a burning passion for this community and have a real love for one another. Through this book they had the opportunity to show *"How [very] good and pleasant it is when brothers live together in unity…The Lord has pronounced his blessing."* (Psalm 133:1,3b)

Our gratitude is "graphically" expressed to Kim Avila for her refreshing creativity in the cover design. She clearly reflects the diamond of devotion to the Lord she serves.

The very idea of this book would probably be found somewhere among the rusty piles of forgotten scrap were it not for our project manager, Rosie Dewey. She served with joy out of her love for the Lord and people, and with a spiritual gift of administration that enabled her to minister so effectively. She grabbed the baton with passion and ran the race to the finish like the champion she is. Well done, good and faithful servant!

Warm and heartfelt thanks are extended to an amazing group of talented individuals who shared our vision and rose to the challenge. They proved we did have the right people with the right skills! Without their dedicated efforts, these inspiring pages would never have been written.

INTERVIEWERS
Marlene Drew
Tina Hansberry
Sam Krahenbuhl
Lisa Sigler
Joanne White
Susan Colt
Starla Romano

WRITERS
Marlene Drew
Melanie Rosecrans
Tina Hansberry
Miriam Greenman
Susan Weis
Janet Bunch
Diane Jasper

PROOFREADERS
Joanne Mills
Jeri Neuberger
Bob Bunch
Lisa Sigler
Jean Roberson
Joanne White
Elaine Darrow
Karen Krahenbuhl
Sam Krahenbuhl

EDITORS
Joanne Mills
Bob Bunch
Judy MacMillan
Ken Ratliff
Letha Ihrke
Bob Walker

Foreword

The people of Cle Elum, Roslyn, and Ronald, Washington, have a dirty history—coal mining. There couldn't be a dirtier job! Black soot and coal dust gets everywhere: inside, outside, and in between. Everything it touches turns black, coal black. The older houses and structures of Upper Kittitas County still retain coal dust in the walls and rafters from the heyday of coal mining, a visible black reminder of yesteryear.

Perhaps your life still carries present emotional reminders of a dark past. A darkness that, even today, clouds the joy of life with regret, sorrow, anger, hopelessness, shame, guilt, or even despair. Perhaps you have wrestled with the nagging, seemingly unanswerable question of what's gone wrong with life. Or even more significantly, you're asking if there is any chance for change, a substantial change that looks and feels like the real thing. If so, we believe that reading these stories of authentic hope will cause that same hope to begin stirring in you.

This simple book contains the stories of individuals (some whom you may actually know) who are all too familiar with life's dark reminders. It tells stories of lives blackened by the soot of everyday struggle, failure, and mishap, only to be transformed by God into a gem of incredible beauty and lasting value. It offers hope, real and lasting hope! No, it's not another typical "how-to" or "self-help" handbook. It is simply a storybook that tells how people, just like you, found a reason to keep going in the face of insurmountable odds, and found the strength to persist. They discovered that a "far away" God wasn't so far away after all, and that He is able to bring reality to what was once only a far-fetched dream.

The hope of these transformational stories can be yours, too. The only difference between you and the people you will be reading about is that they have already made the discovery: coal is the stuff of diamonds. God can transform your life as well!

1

You Choose Who Will Die first

Written by Tina Hansberry

I was twenty-three years old, married for five years, and had no idea how I would get out of the hell I was in. I stood in our living room, positioned and lined up next to my three children, who were all under the age of four. He paced! He snarled! He cursed! The shotgun was waving in his arms. As if briefly possessed, he yelled within inches of my face, "You choose who will die first!" This continued for what seemed like an eternity, but actually lasted about forty-five frightful minutes. This was staged like an execution, but it wasn't going to be a drill this time.

I remember thinking, "This is where it ends; we aren't going to make it out of this one." His quick movement above the vision of the shotgun barrel briefly broke my thoughts. This crazed man had a name, Jonesy. He was the father of our three beautiful children, and supposedly my loving husband. He would walk around behind us, next to us, in front of us, and everywhere else in the room. Looking at him that day, I remember the blackness of his eyes. They were like the pits of hell. My husband wasn't in there. No one was. It did end, and we did make it out of that crazed situation alive. However, there would be more like this one. The anger and torment would not stop until much later in our lives and after many more horrifying situations.

I graduated at the age of seventeen. Catholic school had been my home away from home for twelve years. I was a model child, homecoming queen, cheerleader and honor roll student. School was my comfort and it came easy to me. I could easily be away from home with no concerns about my assignments. Despite my family life, I had made it through and kept focused.

My parents were strong Catholics, so they made their marriage work, regardless of the consequences or who was hurt by the relationship. My father was a wonderful man and really the one that kept us as together as we were. My mother, oh, my dear, dear mother. I remember being a small child and having thoughts that were way too big for me. I remember cringing at even the whisper that I would "be like her." I was scared to death. My mother was certifiably crazy, manic-depressive, bipolar, or maybe even all three. Early memories include my mother breaking a large, hanging wall picture over my father's head because she wasn't fond of his gift to her. And dear ole' mom would often propel coffee canisters though the kitchen that frequently hit my father as he was trying to escape. When she had one of her temper tantrums, my sister and I would scrape Sunday dinner off the wall, just to make sure our tummies were full that night. I remember the time I saw my mother laying on the ground in front of the truck tires, sporting only her nightgown, making certain my father didn't flee the "fit" that was taking place. All this crazed-type behavior primed me to believe that this was not only acceptable, but also normal. I assumed families must be like this because I knew no other, and it was never discussed. As an adult, years later, counseling taught me differently.

So the minute I graduated, I fled to my favorite escape, the horse track. Our family had always been involved with thoroughbreds; it was the one place I felt free and that life was good. It was at the track that I met and fell head over heals in love with Jonesy. He was nineteen and very talented. In spite of his weight struggle, he was one of the best jockeys, called "reins men," on the West Coast. Jonesy was only 5'6" and weighed 135 to 140 pounds soaking wet. The suggested weight for that height was 115 pounds, but he still excelled. Life was wonderful. The track and shed row were home. I couldn't imagine being any other place with any other person.

The signs of abuse started early, but to me they were very normal. His anger and cursing were nothing compared to my mother's, even if they were over "nothing." It was all good. I loved Jonesy and he loved me. During the winters, we would still "break" the two-year-old horses, and fill our time with other fantastic ventures. Fishing, hiking and car trips were part of our passion. Later, our children also became our passion. Life was wonderful. The more profound abuse didn't start until our first child was about one. Then, as the years progressed, the pattern evolved and took on new and more torturous faces.

It was the winter of 1977, and we had rented a place on a small, clear-blue lake that sat high above Mon-

roe, Washington; a place where dreams would be realized, or so I thought. I was home caring for our one-year-
old son and had dinner ready. We were both waiting with
excitement for Jonesy to return from town where he was
breaking in the young horses. Dinner turned cold and the
evening dragged on; Jonesy was not coming. Did I do
something to upset him? I placed our small child into
bed, locked up the house, and turned in for the evening. I
laid in bed worrying for hours and finally drifted off from
exhaustion. I awoke to loud noises coming from downstairs. Jonesy was screaming in a rage, kicking and
pounding on the locked door. After I opened it, he stumbled into the house, obviously drunk. I followed him into
the living room and it happened quickly. Jonesy doubled
up his fist and hit me so fast and hard in the eye that I
saw stars and every other shape in the universe. I was
shocked and dazed. No words were exchanged after the
blow. The only thing left was a closed, black and blue
eye for me, and a need to go to bed and rest for Jonesy.

The next day came with nervous anticipation.
Jonesy was so upset over what he had done, and cried, "It
was the alcohol. It won't happen again." Of course, I
believed him, and Jonesy had me where he wanted me
from that moment on. As I comforted him with my swollen face, the vicious cycle was set. The future was colored with more violent abuse by him, and more comfort
from me to help him not feel so terribly bad for what he
had done. These events should have been enough to
show me the pattern. I should have known there would
be more, but I loved him. I blamed my pain and abuse on

many other things, but never on Jonesy. I thought our life was great. I would mask the swollen, vividly bruised and mangled eye with a concocted story of being poked by a stick while out gathering livestock.

More or less, the violence continued over the years. However, the face blows lessened as Jonesy discovered new targets. Body blows, followed by one-sided kick box matches with me on the floor, took their place. Our violent life, and the marks it left behind, was easy to hide. There was no middle ground any longer. Jonesy was either the best or the very worst. The beatings were always followed with his anger-ridden voice demanding, "Why didn't you just do what I asked you to do?" or "This wouldn't have happened if you had just behaved!" Or better yet, "Why do you have to act like your mother?" Jonesy knew exactly which buttons to push, and he was the master switchboard operator. He flipped switches at will and controlled everything.

For most people who are not used to abuse, the normal stresses of life came and went: Christmas, moving, and other routine events in our journeys. Jonesy, however, had no tools to deal with these life issues, other than what he had learned from his father. He didn't want to be abusive any more than I wanted to be like my mother. Jonesy wasn't strong enough in his fight against being an abuser, and he couldn't escape it. So we walked forward and my focus turned toward protecting our children, now numbering three. I kept the violence from them and took the punishment in silence. Earlier, I learned that silence was my best defense. The children could not hear my screams, and Jonesy was not further fueled by my exclamations of pain.

I had made it to my twenty-third birthday. "Could this be a new season with no abuse?" Often I would sit and ponder how I could make things different. "What could I do differently? How could I be more supportive of my husband?" All of my efforts didn't change the outcome. Yet, I tried, oh, how I tried!

We set out on one of our family trips, this one ending at Jonesy's parents' house. The day started out as one of those great days. Love filled the pickup truck. My youngest child was on my lap with the other two piled between us. Jonesy suddenly spit fear into the air as he asked one of his famous accusatory questions. Prior to answering, I frantically considered my options. "If I answer this way, I am dead. If I answer that way, I am dead. How do I get out of this one?" I answered and sure enough his temper flared. I received a backhanded blow that immediately closed my eye and disfigured its appearance. Quietly we drove the rest of the way. My thoughts were only centered on, "How would I mask this from his parents? Surely they will know. What will I tell them?" But mostly, "Why didn't I just answer him correctly?" We arrived at their home and entered. Immediately, I sensed they knew. They said nothing though. Again, this was acceptable and normal. I was always being told in some way, even by their silence, that this was all okay. I believed it.

The years came and went. We had good times and bad. We now had four children and I had learned by then how to read the signs. We were surviving. My emotions were gone except toward the children. They had lowered themselves to having fistfights with each other at the bus stop. The loser won the prize of having to be the first one through the door on days I wasn't there. The first to arrive always had to run the gamut of Jonesy's discipline and torture. He called it discipline, but it was abuse, plain and simple. The phrase "duck and run" was our motto. We lived by fight or flight.

As Jonesy sensed my growing strength, his power needed a new twist. I considered leaving, and he could see the signs. Attempts of suicide added drama to the aftermath of the more typical events. He sought counseling, on and off, to show his willingness to change. Counseling always resulted in my being counseled about creating a non-threatening home for my husband. They suggested that several minutes of quiet time and loving support, when he arrived home, would help curb the anger that he sometimes showed. Sometimes? Didn't these counselors get it? I tried all that! I wasn't the problem! I couldn't fix him! So, on we went.

I had my first job outside of the track. A friend of ours had a machine fabricating business and I worked in the office. An office party was to be held and Jonesy encouraged me to go. He didn't want to mingle, but he really wanted me to attend. The break would be nice, so I agreed to go. Later, after the party, my husband became enraged and accused me of having an affair. I went upstairs, sat on the bed and began to read. I wasn't doing this again! I refused to play this time. I didn't look up; I just read. The door to our room flew open and the madman came through. Jonesy leaned over the nightstand to pick up a shotgun shell that had been there for years. He grabbed the shotgun from the closet and loaded it. I didn't look up. Jonesy grabbed me and stepped in front of my body. He placed my arms around his back and curled them on his stomach, holding them forcefully. Jonesy placed the butt of the gun on the floor, with the barrel facing upward to his throat. His feet were bare and his toe was balancing gently over the trigger. "You are going to watch me die!" he yelled. Again, my first adrenaline response was to pamper his ego and his feelings. I was finally able to free my hands and push the barrel of the gun forward. Now free, I raced into the bathroom, pulling at my own hair in panic and fear. How was I getting out of this one? My children were downstairs. Do I jump out the window? No, I can't do that, he will get my children. My thoughts were crushed as he grabbed my hair and yanked me back into the bedroom. He had changed his mind. He shouted, "I'm not going to die today, you are!" As he pointed the shotgun at me, Jonesy demanded, "Aren't you going to beg for your life?"

For the first time, I spoke during one of his fits, and screamed, "Go ahead, get it over with. I want to die. I want this to be over."

Jonesy pulled the trigger. Rather than a shot, all I heard was Jonesy yell, "BANG!" The monster had taken the shell out of the gun while I was in the bathroom. I was really MAD. I had wanted it to be over and it wasn't. It was obvious that I had to get out of this situation or someone was going to die. This was not normal. More than fifteen years of violent abuse was enough.

We didn't move out after that near-death episode. In fact, it took another eighteen months and a second shotgun rally before I was finally able to tell the children we were leaving. This was extremely difficult because we all loved Jonesy. The good times were incredible and he was the leader of our family. Nevertheless, I planned to leave and my children were coming with me.

We went to Wyoming for a wrestling tournament, and stopped to visit friends in Spokane on the way home. The plan was to go home, get our things and move on. However, Jonesy sensed this plan and called us in Spokane. His threat of suicide fell on deaf ears this time. I calmly advised him to go ahead and do it. Furthermore, I requested that he not leave a mess for the children and me, and hung up. Cold and callused, maybe, but I couldn't be anything else any longer. He had used me up. Emptiness was all I had left to offer. The kids were in tears and couldn't believe this was the way it would end. Jonesy was their father, and I had just given him my approval to commit suicide.

We continued home, packed up our things and left. On our way, we stopped at the track to check the horses and to say goodbye. When Jonesy saw what we were doing, he flew down shed row with rage on his face. He passed us and went straight to our vehicle, which was loaded with our getaway money and all of our valuable personal items, jumped in and drove off. That was our life. The kids and I knew we had to get the car back to begin our new journey. We headed home to retrieve what little more we could. Jonesy met us in the driveway and began beating me violently, this time in public and in front of the children. I was being punched and kicked with no apparent end in sight. My ten-year-old son showed the only wisdom our family had seen in years. He called 911. Jonesy was picked up, arrested and taken to jail.

He was arrested on Friday, arraigned on Monday, and then kept in a psychiatric ward at Harborview Hospital because of the suicide risk. While in jail, and at the hospital, he was able to keep up the endless calls telling me he would kill us when he got out, and threatening suicide. If I wasn't at home, he would leave the terrorizing message on the answering machine. They released him after just one week. I slept in a chair in front of our door with a shotgun on my lap. I was ready to do whatever was necessary to protect my family. Fortunately for me, Jonesy didn't come back.

Jonesy hated jail. He felt he couldn't handle being locked up, and had a great fear of going back. That was what saved us. I agreed to not press charges if he would agree to leave the area and honor the "No Contact" restraining order. He didn't leave town for another three months, and continued making the terrifying calls to me. When Jonesy was faced with returning to jail for failure to obey the judge's orders, he finally packed up and moved to California. The officers told me, "You're finally safe." There were still many sleepless nights after that, wondering if he would reappear. He didn't.

The next five years were probably the worst. I wish I could say it was all easy after he left, but I would be lying. There was so much healing to be done. I didn't know who I was or how to feel. I remember sitting on my bedroom floor frantically rummaging through old pictures. Maybe I could find just one picture that would remind me of me and what I was about. That picture was never found.

I began intensive counseling with a wonderful woman. She would interject God occasionally, only to have Him immediately rejected by me. I didn't want her God! Look what He had done for me so far! I would cry, thinking, "What kind of God would put my children and me through this horror?" Even in my Catholic upbringing, Jesus only received honorable mention at Easter and Christmas. So this idea of a God or Jesus Christ was hogwash. I started searching for something else, anything else. I experimented with new age, old age and every religion in between.

I moved to Cle Elum, Washington, in 1994, a location central to all the racetracks. I was beginning to heal. I met a wonderful woman that invited me to church. I accepted, but told her that I would only attend after they were done with prayer. So, true to my word, I only entered the church after prayer time. One day, I came in just on the heels of a finished prayer. I asked, "Does this crap really work?" There was humor to my unpolished question and the dialogue began. More seeds were planted. I found it nourishing. I continued this new diet of worship and even attended a women's retreat.

They spoke of forgiveness at this conference. All I could think about was, "How could God, if He existed, forgive me for what I had put myself and my children through?" I wanted to know. I returned to my Cle Elum home and received the answer.

I was standing in the shower, pondering the day's discussion. I felt such deep regret and sorrow for what had transpired in my life up until now. The tears were flowing, and I asked God, "How could You ever forgive me?"

An audible voice answered, "Because you asked, I promise you will never again walk through dark times alone."

I rushed out of the shower, threw a towel around myself and asked, "Who's there?" After a quick look through the house, I realized it was empty, except for the Holy Spirit and me. Stepping back into the shower to rinse, I confess I felt angry. Could it be this simple? Where were the agony, the fasting, and the running naked through a rainstorm to be cleansed by lightning? As the truth began to sink in, the irony became apparent. All my sins of the past, my sorrows and my guilt were washed down the drain along with my tears, just because I had opened my heart to ask for forgiveness, and God had answered. It has been the only time I heard His voice aloud, but it was all I needed for healing to begin. It only took hearing His voice once for me to know that God will always be with me.

That afternoon, God filled my heart with a joy and

peace that I had never known. I am now a complete person. I don't worry anymore, as I have complete trust that God will see me through anything and everything. It may sound morbid or crazy, but I wouldn't change the path that my life has followed. I am grateful for my life today. My many trials have led me to where I am today. I, for sure, wouldn't change that.

I pray about everything in my life, and God has answered each and every time. I have been given a wonderful new husband, and have been healed enough emotionally to enjoy what real love is. Although tragedies have come my way (in the form of my husband falling off our new roof), God has blessed us. My husband shouldn't have walked again, but I prayed and prayed like there was nothing else that mattered. My husband is healthy and whole. My daughter had a child with severe birth defects, but that child is still alive, through prayer. God will prepare me for everything that comes my way. The difference today, versus twenty years ago, is that I accept and believe in my Savior. I am alive and have peace. I love God and need nothing else. Recognizing that God is forever by my side is strength beyond description.

2
A Great Man Bearing Gifts–Huckleberries

The Story of Kathy Juris
Written by Marlene Drew

"Kath, be sure and take your cell phone today." Simple little words from my husband, Bill, as he walked out the door on his way to work. What would I have done if I'd known these were the last words he would ever speak to me? Would I have begged him to stay home? Would I have stayed with him and locked all the doors and windows to keep life from happening?

It was November 14, 1995, just another day in a very satisfactory life. My husband and I had already done the ritual hug, kiss and "see you later," a routine I loved and one that would take on such monumental importance. It was a defining moment in my life, a literal cornerstone in my approach to future relationships.

My mother and I were taking my brother, Bob, across Snoqualmie Pass to SeaTac Airport so he could head home to Alaska. He had been visiting the family and it was the day he was leaving. After dropping him off at the airport, mom and I were going on to Tacoma to visit with mom's sister, Aunt Lois, and to meet my daughter, Linnea, for lunch. It was a great visit with everyone, lots of laughter and easy conversation. We are all very close and have never had a shortage of words between us. Soon, mom and I had to think about heading home in time for yet another family occasion. My grandson, Alex, was turning seven and we were having a family birthday dinner at 5:30 that evening.

While on our way home, I checked the time and realized we were running a little late. I had my new cell phone with me, having just gotten it the day before. Since I wasn't sure how to use it yet, I pulled over onto the wide shoulder beside Lake Keechelas to call Bill to let him know our arrival time. I punched in the number to my house, and feeling quite accomplished, pushed the send button and listened for the ring: one…two…three…four…five. No one picked up the phone! I thought it a bit unusual. Bill should be home from work by now; he is always so punctual. I wondered where he was, but didn't feel too concerned. I dialed my daughter's house and got no answer. Now this was a puzzle since we were to have dinner there in just a few minutes. "Where is everyone?" I wondered aloud to mom.

She said, "Let's call my house."

I dialed my parents' home and finally a voice–my dad's! Right away I sensed something was wrong. His tone sent a dreaded chill up my spine. "Honey, come home…there has been an accident!" Those words echoed around in my brain: An accident…an accident…an accident!

"Who is it, Dad?" I screamed. "Who…what? Dad, please tell me!"

He repeated, "Honey, just come home." My heart climbed into my throat, pounding so hard I thought I would be sick right there. Every person in my family flashed into my thoughts. Which one of them was hurt?

As I pulled the car back onto the road to drive those last few miles home, I envisioned a dozen scenarios. My little grandson was just learning to ride his bike, and there was a sharp curve in the road where we lived. Could it be him? Who…what…how bad? I looked at my mom and her hands were clasped tightly in prayer. She was praying to the Lord as hard as she could as her mind was racing through the same thoughts as mine. I was trying to keep the car between the lines, wondering if I could actually drive us home. I started sweating a sick, soaking sweat. My hands were slick with it, and it was also running down my back. I was praying for God to just help me make it home.

As I came down the hill towards Easton, just a few short miles from my home, I saw several State Patrol cars with flashing lights, an aid car, and several other vehicles. No, No, No! My mind would not go there; I did not want to think that any of what was going on there had something to do with me. I purposefully did not look, but kept my eyes and my thoughts on the task at hand—just getting home so we could find out what had happened to one of my beloved family members.

Cars were everywhere when we drove into my parents' driveway. I knew someone must have died. My legs could barely carry me to the door. I wanted to know what had happened, but at the same time, I was scared to find out. The knowledge would make it real and there would be no going back. Dad met us at the door, put his arms around me and said, "Kathy, I am so sorry to tell you this, but Bill was in a car accident and he died." I couldn't quite grasp what he was saying. Was he telling me that Bill, my Bill, was dead? Surely not! That very morning we had said, "See you later." I took my new cell phone with me so we could stay in touch I hadn't even talked to him on it yet! There were at least fifty of our family and friends gathered in the room, but someone was missing. It was Bill.

My dad, as gently as possible, explained what had happened. It seems that a young man, and his even younger girlfriend, had been on a crime spree over on the west side of the mountains. They had been stealing cars, and for some reason, had decided to bring their antics over to our side of the mountain. They were in one of the stolen vehicles, driving erratically across I-90. A number of other motorists had already notified the State Patrol of the reckless way the car was being driven. Bill was the manager of the ski resort on Snoqualmie Pass. He and his crew had finished their shifts and were heading down the mountain, homeward bound. Bill was only a few short miles from our home, when the young man rammed him from behind and pushed him off the road. He was killed instantly. Members of his crew, who saw the accident, were the first on the scene to help. The young couple also stopped, but only long enough to steal one of the Good Samaritan's vehicles. They were apprehended later and charged with vehicular homicide and auto theft. This just did not make any sense! It was a classic case of being in the wrong place at the wrong time, tragically so. It had all the makings of a badly produced Hollywood movie.

As dad finished telling me the story, a sickening thought came to me. That was Bill—the wreck I had driven past on my way home! My husband had died in that wreck and I had driven right past it, not knowing. How could I not have known? I went numb all over.

Bill and I met in high school and started dating our sophomore year. I lived in Cle Elum, and he was from Roslyn. We both came from long-established families in the area. We always knew we would marry, and we did soon after we graduated. We stayed in the area to be near our families, bought a house, and began our lives. Bill was a good husband, a good father, and an excellent provider. We had a good marriage, not one lived in a state of animated bliss. It was a real union of two people who came up against the normal problems of living as man and wife. The crucible was that we loved each other enough to work out any difficulty that came our way, and had a wonderful relationship. We were looking forward to our golden years together, enjoying our parents, children, grandchildren, and our extended family and many friends. It seems all of that changed in an instant. I knew that my faith in God was going to be tested, and I wondered how I would handle it.

The next few weeks were filled with incredible sadness and a pervasive fear. How could I go on without Bill? After 35 years of marriage, how would I make it without him? Could I take care of myself? He had always taken care of everything—the household repairs, the vehicles, the lawn–everything. I wasn't sure if I could even start the lawnmower, let alone take care of the house.

Every time I became overwhelmed with sadness or despair, the Lord would send someone to comfort me. Countless prayers, hugs, phone calls, gifts of food, words of encouragement, and even financial support surrounded our family. I was never alone; my daughter and her family from Arizona stayed with me throughout the winter. Our many friends were always available, and my church continuously lifted us up in prayer. I believe God puts people in our lives for a reason. At any given moment, someone would deliver the exact words of comfort, right when I needed it the most.

I had always been raised in the church, and our daughters were given this same spiritual upbringing. We had a solid foundation. I knew who my Savior was and I knew He loved me. Throughout my life, faith was always there, sometimes stronger, sometimes weaker, but always a constant part of who I was. Throughout that long, lonely winter, I learned that I had taken faith for granted, just like people who take their perfect health for granted and then are stricken with a terrible illness. All of a sudden, you are put in the position of having your faith tested, not in just the day-to-day little things, but in ways that have enormous ramifications.

I had seen people who, upon having lost a loved one, curled up in a ball and literally could not go on. I knew God wanted more than that from me. Despite being surrounded by wonderful people, the reality that Bill was gone was sinking in, and I was having some very lonely days. Certain little things, a song, a verse, a perfect sunrise, would trigger a sense of deep loneliness. I remember one day in particular, sometime around Easter. Bill had been gone for about five months, and I hit an all-time low. I got in my car and just drove. Ending up in Ellensburg, I wandered around the back roads, crying, thinking, praying, and asking the Lord for strength. I felt His presence and I knew that I was going to be alright. I knew that the Lord had a plan for me and He was asking me to trust Him, just as I had always done. I knew I was going to make it through the test that had been put before me. I thanked Him for the life that He had given me, for the wonderful years with Bill, for my beautiful daughters, for the blessing of my grandchildren, and for the many friends and people that had been placed in my life. But most of all, I was beginning to see a message in the midst of the tragedy: to never panic when something bad happens, because God always has something for you just around the corner.

Content with the knowledge that I had already experienced the gift of a wonderful marriage, I decided that dating was for the younger set. The thought of another man coming into my life was as far-fetched as it could possibly be. But God brought me Stu. He lived just two miles from my house on the same road. We knew each other as neighbors. He had taught my daughters in school, and my girls had babysat his son. Before Bill died, Stu's wife had passed away from cancer. I remember driving past his house on my way to work and wondering how he was coping with being alone. I would see Stu during the week at the supermarket where I worked, when he would still bring his in-laws to the store to do their weekly shopping. That act of kindness had always touched me and showed such great character. I worked in the meat department, and they would always come by to get some bones for the dog. We would exchange greetings and a few pleasantries–at least, that's what it was on my part. I did notice, after a time, that the dog seemed to need a lot more bones and our conversations were lasting a little longer. I had no idea that the bones were piling up at Stu's house, and I was even more in the dark about what my family was doing behind the scenes.

It seems that our family and friends were determined to get us together. One day, my son-in-law informed me that there was going to be a wiener roast, and I needed to be there. I told him I was just too busy at the time. He let me know that this really wasn't an invitation, it was a requirement. He had already invited Stu, and was frantically trying to round up other family members and friends, so this wouldn't look like the set-up that it actually was. Normally, you can make one phone call to our family members and get together a good-sized group. This one crucial time, everyone was busy. He did manage, however, to pull enough people together to cover the fact that this was, in actuality, a matchmaking scheme. I attended, clueless to all the devious planning, and had a wonderful time.

A short time later, Stu invited me to go huckleberry picking with him. I had an appointment that I couldn't get out of and declined. I really wanted to go, and was surprised at the disappointment I felt. Later that evening, I saw a car pulling into my driveway, and it struck me that this was living proof of what I had learned during my time of deep sorrow. "God always has something for you just around the corner." It was coming to me in the form of a great man bearing gifts—huckleberries! I took the berries, made a big pie, and delivered it back to Stu. We were married two years later. The families on both sides couldn't have been happier; no more plotting, no more planning. They were thrilled, and so was I!

Never give up on God; He loves you! He will always give you what you need, maybe not exactly as you want it, but what you need. Never miss the chance to reach out and help someone with kind words and kind deeds. These have lasting effects on others, more than any of us realize. Never take your faith for granted, but treat it like a precious jewel you wear with reverence, not one you keep in a box. And last, but not least, never let anger or resentment go unresolved. I cannot tell you how thankful I have been that the last words that passed between Bill and me had nothing in them but love.

3

A Healed Family

Written by Marlene Drew

JESSICA:

I was six years old, and I didn't know any better. I was daddy's girl. He loved me, and I thought that was how he showed it. That is what he told me and I believed him; he is my dad.

That is how it was for me when I was growing up. My dad had a temper, and he knew how to lose it. I knew that after our "special times" together, his anger would abate for a while, and he said I meant so much to him. Our time was so important because his life was stressful, and I was daddy's special girl. I never talked to my mom about this. Dad told me not to. He said she wouldn't understand about our special bond, and she would be mad at me because she might think he loved me more. I was afraid she would blame me for daddy loving me too much, and I didn't want her to be angry. How could I go to her with this? Somehow, even in my innocent young mind, I knew this would ruin all our lives. So I endured; I curled up inside of me, and I endured. This went on until I was thirteen years old.

We had been living in Montana. Dad got a better paying job in California, so our family moved when I was about ten. I made new friends, more worldly friends, and became aware that the relationship between my dad and me was not the same as what the other girls had. I started to run wild, walking the streets, drinking with my friends, and hanging out. I didn't want to be in the house anymore. The mother of one of my friends was on drugs and drinking, so I stayed over there a lot. We didn't have to sneak out; we could just walk right out the door and no one would notice. We would bum around on the streets or go with her brother, who was sixteen, and hang out with his friends. When my parents figured out what I was up to, they told me I couldn't go, so I snuck out the window and went anyway. Sometimes, I would just climb out and sit in my yard. It wasn't about going anywhere as much as it was about not being in the house. As for boys, I was a tease. In my warped sense of right and wrong, I enjoyed the kissing part and the hugging, which felt good to me; the scary, painful parts, I got at home.

DAVID:

From the beginning, all I knew was abuse. My own father was physically, verbally, and sexually abusive. Anything could, and would, set him off at any time. All eight of us kids, five girls and three boys, lived by "walking on eggshells." Dinner was never an easy time for any of us. Dad sat at the head of the table and watched, just waiting for someone to mess up. We were supposed to eat anything put before us, and act as though we liked it. If we didn't, those lightening-quick fists would send someone flying. At every meal, my throat would close in fear. I could hardly swallow my food, and it would get stuck halfway down and sit there like a lump.

My father molested each one of my sisters, all the while they lived at home. At first, I wasn't aware of what was going on. It took until I was about age six to figure out that the something evil lurking in the halls at night was my old man.

That same year, some neighborhood boys grabbed me and drug me into their house. No parents were there. They molested and tried to rape me. I managed to get away and ran home, but didn't tell anyone. What would have been the point? In my world, this was just another day, not unlike what went on in my own house. Who would I have told? Who would care? I'm not sure if I even understood that this was wrong; I just knew that I didn't like it. It made me really angry that they would do that to me. I wasn't sure if I lived in the worst place in the world, or if it was like this everywhere. I had no way to really judge what was right from wrong, what was natural or unnatural. We were sure my mother knew what was happening, but chose the role of denial and did nothing. Was it right? Was it wrong?

By the age of eleven or twelve, my brothers and I began to understand that what was going on in our home wasn't normal. My sisters had started to communicate with each other and with us about what was happening to them. Up until then, they didn't know that it was happening to the others as well. We all figured out that if the boys stuck around the house whenever my sisters were at home, the old man wouldn't do anything to them. We were just kids protecting the girls, which should have been his job, not ours. I started carrying a gun about then, feeling I could be a better protector if I were armed. As the years went on, my anger grew and I couldn't wait to get out of the house. After graduation from high school, I joined the Navy, and married my high school sweetheart. That lasted about six months, and her rejection of me just added to my anger.

SUZI:

I came from a family of eleven children. I was one of a set of twins, born smack dab in the middle of the brood. My family was Christian and had a solid church foundation. My parents fought and argued, but it was nothing compared to David's folks. I led a relatively normal young life, no big events and nothing earth shattering.

Sometime in my junior year, my brother introduced me to David and we started dating. He seemed like a good catch. He was four years older, in the Navy, heading to California, and could get me out of the house. David wasn't as eager to marry as I was, since he had been through it before and it didn't work out for him. But I was persistent and kept up a correspondence through the mail. We got to know each other better that way, and he always wrote back. He could write easier about his feelings than he could voice them. We were married during my senior year of high school, and I was able to graduate early, since all of my required courses were completed. I was so happy to be leaving that house full of people and getting out of the State of Washington. It was an exciting time for me.

DAVID and SUZI:

We married in March of 1975, moved to a Naval Base in California, and lived there about two years. We stopped going to church and picked up drinking along the way. Jessica was born in 1976. About the same time that David was getting out of the service, we moved back to his hometown. It was a real "hick" town, with about one hundred kids in the high school. We liked being a family with a baby, and thought we were ready to settle down and do all the right things. David got a job that required a lot of long hours, but didn't pay very well. When the financial stress hit us, some of David's past started to emerge.

He shared with me the things that happened in his house growing up, and about what his dad had done to the family. We had not been living there very long, when David's dad showed up, unannounced, early one morning. Under the pretext of bringing in the mail, he walked right into the bedroom where I was still in bed. I kicked my foot at him and yelled, "Get out!" The doors were never left unlocked after that, and he wasn't allowed in our home unless David was there. David guarded Jessica and Suzi from all the pains of his past, but they weren't over yet.

David had to travel in his job, and it broadened our horizons a bit to the wide world that was out there. It became obvious that there was no future in this small town. We just couldn't get ahead. We moved to Montana to make a better living and have a nicer home.

Having been trained for electrical work while in

the Navy, David found a pretty decent job. The one thing it didn't provide for, however, was medical insurance. Our second baby was born and contracted spinal meningitis. The hospital bills were huge, and this added more pressure to the already-strained finances. We still weren't going to church. Our drinking continued to increase, and David's temper started to model his dad's.

DAVID:

I just couldn't seem to get anything right. I really was excited about being a father and a good provider. Every time it seemed we might make headway, something would come up. I could feel the anger growing in me. My temper would flair at little things, and I would lash out at anyone in my path. Alcohol seemed to soothe the beast for a little while. I remember a time that I told God I was going to do my own thing. I felt sure that believing in God worked for some people, but not me. The longer my attitude remained that way, the worse my anger became. It was a downward spiral, and the boiling point got closer and closer to the surface. Out of anger, I hit Jessica once because she was crying about something, but mostly because she was just there. I screamed at Suzi whenever something she did irritated me. I would yell and carry on if anything bothered me. It was during this time that I started abusing Jessica sexually, when no one was around. Suzi would frequently be out of the house, and I would take Jessica to our room. After it was over, I would feel so horrible. I would lie awake all night crying and asking myself why I had done it. It killed me emotionally. I would vow to never do it again. That might last a little while, but sometimes it happened again right away.

I found a good-paying job in California and moved the family there, where we started to get out of debt. We were able to pay off a lot of the hospital bills. The abuse was still going on with Jessica. I felt like I had gotten into a cycle and didn't know how to get out. My drinking got worse, partly because of the abuse, and the abuse got worse, because of my drinking. I drank to get rid of the awful feelings of remorse. I felt terrible from the perpetrator's point of view, and I can't imagine how it was for Jessica. I was putting Suzi through hell too, and she knew nothing about what Jessica was suffering.

SUZI:

I didn't have a clue about what was happening with Jessica. I knew that David's drinking had gotten worse. It was literally out of control. When he would blow up, I usually gave him another drink to try to calm him down. It was obvious that our marriage was in jeopardy, but I had no idea how bad it would get.

I thought I had always protected my kids from the things that could harm them out there in the big, bad world. I really thought that my husband and I were of one mind when it came to that. Jessica had endured six years of abuse. I had been too busy with my own problems in our marriage to read any signs of what was going on right under my nose.

JESSICA:

God sends angels in many shapes and forms. Mine was my aunt, who moved to our town. She lived in a house near my school. Every afternoon, the minute school let out, I would go over there and spend time with her. Because she was family, my parents never told me that I couldn't see her. I had a safe place to go, and I didn't have to go home. I really think she knew something wasn't right at my house, but I never offered to tell her, and she didn't ask. What she did do was to take my sister, my brother and me to church. We were overwhelmed the first time we walked in the door. People hugged and welcomed us, and made us feel cared for. I've never forgotten that feeling. It was so strange to me, to feel so much love in one room. Because we loved the atmosphere there, I wanted my parents to share it with us. I bugged them so much that they finally went with us a couple of times. But usually they would say, "We'll go another day."

A guest speaker was coming to the church to give a seminar on "Date Rape," and I decided to go. The associate pastor, knowing nothing about our family life, told my parents that she felt they should attend such a topic with me. Mom and dad agreed to be there with me to listen and see what it was all about. The speaker spoke straight to my dad's heart. He said, "If you are an abuser, you can never get out of the cycle without help–the kind of help you get from God." He spoke of the power of God's grace and love, and of forgiveness for all sins. I can't believe how powerfully it hit dad. It knocked him to his knees. It was God at work!

DAVID:

I am not sure what prompted me to go to the seminar, except for God at work in my life. With the way I was living, who would have thought He would care about someone like me. Suzi and I had gone to church a couple of times, more to hush the kids than anything else. They were so insistent in wanting us to share what they had found. So I went to the seminar, listened, and heard the most amazing message, like it had been crafted specifically for me. I loved the way the man spoke. I knew after hearing him that my answer would be found in the Lord. The message was so powerful and clear to me. I had never wanted to do the horrible things that my dad had done to my sisters. I did not want that for my kids; yet, here I was, doing the same awful things to the child I loved so much. I knew it would take something I didn't have within me to get me to stop what I was doing. I also knew, right then and there, that what I needed was God. I found the Lord that night.

I went to the pastor and told him what I was doing to my daughter. We prayed together, and I asked God for forgiveness. The feeling that came into my heart was indescribable. I had been living under my sin for so long that it had become an unbearable weight. I felt it lift off me, and it was replaced with a feeling of hope and joy. For the first time in my life, I took a breath of sweet, clean air. The pastor and I set up a meeting so we could talk again. I was willing to do whatever I needed to make this as right as I could, even if it meant going to jail. We arranged for counseling with one of the counselors from the seminar. It was decided that I should go to the authorities, and the pastor would accompany me. I had to tell Suzi before any of that could happen. I dreaded the thought. I knew this would hurt her, and that was something I never wanted to do. I wasn't sure if she would even stay with me, once she knew what I had done. I really had no right to expect anything from her.

SUZI:

I attended the "Date Rape" seminar with David and the kids. I must say that I was surprised at the way it affected my husband. He was overcome with the Holy Spirit and was on his knees. David talked to the pastor afterwards. He didn't tell me what they talked about, but he did say he wanted us to start going to church. I wasn't as keen on the idea as he was, but it didn't take long to convince me that it would be good for us as a family. The kids were so excited.

A couple of days later, David had an appointment with the pastor. I didn't go with him, and was sitting at home when I saw them drive up together. As I watched them come up the walk, I was clueless about what was going to happen. I remember wondering what they were doing here, and why David had such a lost look on his face. After they came in and we were sitting down, the pastor said, "We have some things we need to talk about. I'd like us to pray first and prepare our hearts to be able to listen to God and to each other." He said that David had come to him and confessed some things to him, and now, David needed to tell me. We prayed, and I sat, waiting, as David began to speak.

As he unraveled the story about the abuse that had been taking place for so many years, my world stopped. There was no longer any air in my universe; my lungs couldn't find any sustenance, and everything began to spin around. I could not comprehend what he was telling me. He had done WHAT to my daughter? His daughter! My stomach lurched, and I was going to be sick. I jumped up and started shouting over and over, "No! No! No!" We were supposed to protect our kids from this kind of thing, not bring it to them in the name of fatherly love. I couldn't feel anything for a few moments, but as the numbness began to wear off, anger quickly rose up in its place. I became furious with him. I wanted to pound him, slap him, and at the same time, I wanted him to comfort me. He was my husband; he was supposed to be there for me in times like this. The questions started to pour out when I found my voice. "How could you? Why would you? You are just like your father!" I remember saying those words. I remember the look on his face as I said them, and the hurt that registered in his eyes. But it felt good, at that moment, to hurt him in any way I could. I told him that I wanted him out of the house. I wondered what I would say to the kids. How would I, or could I, apologize and make this okay for Jessica? How would I help her? Those were my first concerns at that time. My marriage was something I wasn't thinking about at all. I went to Jessica and hugged her and told her over and over and over, "I am so sorry." I didn't know what else to say. I wasted no time going to the hardware store to buy locks and put them on all the bedroom doors. I knew I wasn't going to leave him alone with the kids again. I really had no idea how we were going to make it through this trial, and was just trying to hold on to my sanity.

One thing I realized many years later is that I never thought about divorce during any of that crisis. I had no idea how we would reunite as a family, but neither did I feel compelled to divorce David. Family dynamics had changed, but I still loved him. That hadn't changed.

DAVID:

I started out by telling Suzi about the abuse I had suffered as a child. She already knew some of it, and I wasn't trying to use it as an excuse. Then I told her what I had done to Jessica. I watched as a number of emotions moved across her face, finally settling on rage. She was so devastated and incredibly hurt by my actions. I wanted to comfort her, but I was the one who had caused her the pain. I know the abuse to Jessica was terrible and wrong. I can't tell you why I started abusing her. Believe me, I have thought about it a lot. I could only tell my family that I was willing to do anything necessary to re-pair the pain I had caused, and I wanted to make things right with Jessica

Pastor and I had already talked about my moving out. We had made plans that included my going through counseling and also going to the authorities. None of us had any clue as to what might happen after that. I hooked our travel trailer to the pickup, and drove it to the job site, where I was working. That was to be my home for the time being.

Suzi agreed to go to counseling, and take the kids as well. I could only pray that she would agree to work on saving our marriage. I had no right to ask or expect anything. I only had the power of prayer, and I used it.

JESSICA:

I had no idea what my dad had planned to do, and I can honestly tell you that it didn't register with me. All of a sudden, our lives were in a whirlwind of activities. Dad was moving out. Mom was always crying and mad at him, and she kept apologizing to me over and over. I couldn't figure out why. A few days later, we were in the attorney's office. As my deposition was being taken, I finally realized why I was there and it took me by complete surprise. I was so relieved that my dad was the one who had done the telling. I had read some self-help books on sexual abuse and all of them said the same thing–in the mom's eyes, the daughter will be the one to blame. I didn't want to ruin our relationship. This had been a secret for so long, and I had not told anyone, especially my mom. Because I was so young, I was scared that she would get mad and think that I had done something wrong. I thought she might throw me out of the house. Instead, she kept telling me over and over how sorry she was.

My dad moved out, although I didn't want him to, and neither did my sister or brother, even after they knew what he had done. He was our dad and we loved him. Mom and the pastor let me choose whom I wanted to talk to about what had happened to me. I chose our pastor and the associate pastor. They are the only people that I have given any details to about the abuse. Having them there to pray and listen helped me more than the counselor that I saw every week. Maybe it was because I had already established a relationship with them, but it was easier for me to talk to them about my feelings than to anyone else.

SUZI:

I was struggling with my emotions. My trust was gone, and I didn't know what was real. The only people I felt comfortable around were the pastor and the counselor. The only place my trust felt safe was in God. I could see how the Lord made sure that we were covered before this horrible secret came out. All the right people that could help us had been in place beforehand.

We had gone to the attorney's office and stated our depositions, and had to go before the judge to see what punishment he would impose. The judge was impressed with the counseling that David, our pastor, and I had already begun, and did not think that jail time would serve any purpose. He didn't allow David to come back into the home, and he wasn't even allowed visitation rights. David also had to continue with his counseling for the next three years.

We hadn't been in counseling for very long, when I knew that somehow our marriage was going to survive.

I had never believed in divorce; my own parents separated after thirty-five years of marriage, and I still can't figure out why. After that long, you should be able to work through anything. I believe to this day that our marriage was saved partly due to the fact that we were new to the faith, and that the "fire" of the Holy Spirit was burning hot in each and every one of us. The Lord was going to give us the strength to see this through.

DAVID:

The grace that God showered on me was unbelievable. I missed my family terribly, and I kept on hoping that, through the Lord, we would be able to reconcile. My wife and I talked regularly. By now, we knew we were going to make it, but I still had no contact with any of the kids. The judge wouldn't allow it. In the hopes that someday I might have my family back, I complied with whatever he wanted from me.

I had been out of the house for two years without seeing my kids, when Suzi and I decided that we would renew our vows on our fifteenth wedding anniversary. The judge gave his approval for the kids to be at the church. My emotions were all over the board. I didn't know how Jessica would be towards me; I could only hope that someday she would find a way to forgive me.

JESSICA:

During my sessions in therapy, my counselor seemed continually frustrated with me. In his professional view, he didn't think I was exhibiting the anger towards my dad that would help me heal. I tried to explain that I didn't feel that way. I loved my dad, and had been praying with the pastors to be able to have him back in my life. My time spent with the pastors in my church was different. They seemed to understand that, because he was still my dad, I loved him and wanted our family back together. I learned so much from them and the other men in church. They were all father figures to me. It was during those interactions that I learned what a relationship between a father and daughter should really be. I learned what was proper between women and men, and how to relate to men in a healthy way. When my dad had turned himself in to the police for his crimes against me, I understood the depths of his remorse and had prayed that nothing would happen to him. I thought that if I had forgiven him, then who else should have the right to say anything? That should be the end of it. God had granted me the peace of knowing that all was forgiven.

My mom and dad were going to be remarried on their fifteenth wedding anniversary. I was going to get to see my dad for the first time in over two years. I was both nervous and excited at the same time. After all, he was still my father and I had never stopped caring about him. I didn't know how my dad was going to react towards me, because it had been so long since we had seen each other. I knew where I was in my heart, and I prayed that he was in the same place. I was worried that it would seem weird seeing him after so long. I didn't want it to be that way, and I was hoping that it wouldn't be strained or uncomfortable.

I walked into the church, and there he was...standing up front waiting for us to arrive. I could see the anxious look on his face. I knew exactly how he was feeling, because I had been feeling it too. I was so excited to see him. I knew right then that everything was going to be fine, and the only thought I had was, "There he is–there's my dad." In my heart, everything was over; our relationship was healed. Six months later, the judge let him come back home to live. We picked up our lives, but not exactly where we had left off. Now, I had a dad who was letting the Lord lead the way, and he truly loved and believed in Him.

Looking back, I realize what a sense of peace and liberty I had received when I gave my life to the Lord. I am a grown woman, having been married for twelve years now, and have children of my own. My dad and I have a wonderful relationship. I found that we were able to put the past in the past and leave it there. In many ways, my dad became my hero when he stood up and told the truth, knowing he could go to jail. He didn't have to turn himself in *because* of me, he did it *for* me. I look at the courage it took and have a great appreciation for what it cost him to do it. We are a whole and healed family now. All of us know and serve the Lord, praising Him for the miracle He worked in our lives.

DAVID:

In weighing my past against what my life has become today, some things are crystal clear. If your life is not living up to your expectations, whatever might be happening, there is only one answer: call upon the name of Jesus Christ, and stick with Him. I am amazed with the work the Lord has done in my family. My kids are strong in their relationships with the Lord. My marriage to Suzi has just grown better and better over the years.

The best testimony a man can have is to be able to say that he has been connected all of his life to the Lord. But if that isn't the case, the faster you can say, "I got connected to the Lord," the happier and more fulfilled your life will be.

Hiding abuse is hiding sin. When you are harboring that horrible secret, you cannot live day-to-day feeling good about life. It will eat at you, damaging every relationship you have. You will never know how to truly love again. Give in. Find a pastor you can confide in. Turn away from that sin; ask for forgiveness from God, as well as from those who have been hurt. As soon as you do, a huge burden will be lifted. Put your trust in the Lord. The only way to heal is through the love of Jesus in your life. If you are involved in abuse, or know it's going on, you must get help or give help. There is no other way to break this! What greater help than the power of God could there be to guide you through your life?

4

My Addictions: From Drugs to Jesus

Written by Melanie Rosecrans

The desire, the hunger, the craving, they're always consuming my mind. At the end of the workday when I haven't had anything in hours, I'm desperate, the kind of desperation that's always there. I know where I have to go, the same place I go every time I run out of crack. My car is waiting after my eight-hour day at Costco to immediately drive to that familiar dope house. I don't hesitate. It's late and quite dark, making that area of town even more bleak and eerie. I pull up in front of the broken-down, shady-looking house. The lawn surrounding it is brown, filled with beer bottles, dirty needles and used drug paraphernalia. The windows are covered with sheets and towels so there's not a crack of light allowed in. A few of the windows are broken and the paint is chipped and peeling. It's an awful sight, but it's one of many on this block. Although I'm scared to death each time I go into this miserable building, the thought of being able to finally subdue my craving overwhelms me with excitement.

My dealer greets me at the door. His hair is dark and greasy, his face grimy and his teeth are rotting. Punctuated by dark, purple bags underneath, his eyes are glazed over. The smell of him, putrid and soiled, aided by the tattered and stained clothes he is wearing, is overpowering. He tilts his head toward a door across the room, signaling me to follow him there. We move slowly through this dark, filthy, horrid-smelling house with my thoughts centered on the treasure I'm about to feast upon. The motionless, strung-out bodies draped over the furniture and sprawled out on the floor don't even faze me. Some are unconscious; others are just quivering and shaking, coming down from their high. I realize this is an awful sight, but also know that will be me in a few hours.

We finally enter the room where my dealer keeps his stash. I hand over a wad of money. In return, he hands me a baggie of pure white bliss. I smile and turn to head back to the living room where the other wretched people are strewn. I sit on a stained, brown, torn-up couch, and open my baggie to extract the rock inside. I put it in my pipe, light it up and suck in as deep a hit as I can. "Ahhhhh." It's a moan and a bone-deep sigh of relief at the same time. Slowly, my head leans back against the back of the couch, as I take in that euphoric feeling I love so much. Soon, I lean forward to take another hit, finally satisfying the pestering craving that has haunted me all day.

Buzzing, I stand up and wander around the house. I know a few people, but they're all so high they're unable to notice me. Another hit from my pipe and I'm feeling great now. I head to the bathroom. Opening the door, I see another girl sitting on the toilet, holding a large crack pipe. She's young, like me, and totally strung out. She takes a huge hit, lets out a breath and then takes another. After that hit, she slouches back on the toilet seat and slowly turns toward me. Then I see it: her huge, round stomach. She's about nine months pregnant. A tear runs down my face, but I just turn and leave the room.

The next day, I again go back to that vile house. The pregnant girl has gone into labor and none of us knows what to do. We are all so high we don't know where to take her. At a loss, I lead her into the bathroom past chemicals, drug paraphernalia and rotting food. She delivers that tiny, innocent baby in the bathtub of that disgusting house. I begin crying and start feeling very sick to my stomach. What a horrible person I am. This is hell, pure hell on earth. How did I get to this place in my life? I used to be such a good person. I pace back and forth for a while then go back to the living room, sit down and take another hit. Slumping down in the sofa, I close my eyes and reflect back on my life before drugs dominated it.

At seven years old, the beautiful music of worship swept me up as it chimed through my church. There was a concert that night and the songs were so uplifting and powerful. The theme of the concert was salvation, and the man asked the question, "Do you know where you're going?" I had known about God, Jesus and Heaven since I was a little girl. I think the first time I acknowledged God was right after my mother died, when I was only three. My mom and dad were both Christians, but my mom got heavily into drugs and alcohol, and pulled away from the Lord. That's what ended up taking her life. She died from an alcohol overdose, which is legally a suicide. She left my father to raise my two older brothers and me.

After mom died, my father continued taking my brothers and me to church every weekend. He was very committed to teaching us about the Lord. I was already a faithful Christian when dad took us to a concert. Sitting next to my father that night, I remember thinking, "I'm not sure if I know where I'm going." I told my dad that I wasn't sure if I had Jesus in my heart.

He looked down at me and said, "Well, you need to know that, it's really important." They had an altar call and I knew I needed to go up to the front of the church. I needed to know that I was going to Heaven and that Jesus was my Lord. I asked my dad if he would go up to the altar with me. He took my hand and we walked up to the front. It was there, at that moment with my father and my pastor, I accepted the Lord. I prayed and asked Jesus into my heart.

My pastor asked me if I wanted to be baptized. I started crying and said, "Of course I do. Isn't that what I'm suppose to do?" Since I was only seven, I was somewhat mistaken about what being baptized meant. When I had decided that I was going to commit my life to Jesus, I actually thought it meant I was giving up my life on earth to be with Him in Heaven. So, when the pastor asked me that question I cried, thinking about the life I was giving up. I was still holding my father's hand and I looked up at him and said, "I'm going to miss you." Neither one of them understood what thoughts were going through my mind.

The baptism wasn't for another couple of weeks, so during that time, I was preparing to go to Heaven. I wanted to spend all my time with my father and my brothers. I wanted to go to church all the time. I was ready. I was ready to give up everything in the world to be able to join Jesus in Heaven. The night of the baptismal service, I asked my father if I had to pack a bag. He thought I meant, "Did I need to bring an extra set of clothes to change into after I got all wet?" I actually thought we were packing a bag for me to take to Heaven. So, with my bag of clothes in hand, we got in the car to drive to the church. The feelings flowing through my body were incredible. I was a little upset that I had to leave my father and my family, but I was so excited to finally meet Jesus. My faith was so passionate and so powerful. I had the faith and heart of a child, pure and simple.

As we drove to church that evening, I reached over, grabbed my father's hand and told him how excited I was to meet Jesus. He turned and looked at me and said, "Well, I am too, but I don't think it's going to be for a little while."

Then I asked, "Dad, have you been baptized?"

"Of course I have," he answered.

"Well, how come you didn't go up to Heaven?" I questioned him.

He just looked at me a little confused and replied, "Well, you don't go to Heaven when you're baptized. You are just showing everyone that you've committed your life to Jesus. It's an outward symbol of your love for Him."

All I kept thinking about was all the other people

that I had watched get baptized. The baptismal tank was in the front of the church, allowing for a full immersion baptism. The pastor would dunk them under the water, and then they would go out a different door behind the baptismal. That was the last I saw of them that day. To my mind, that just meant there were stairs up to Heaven behind that door, and that's where everyone went after they were baptized. So I asked him, "Then where do all the people go after they are baptized? Aren't I going to Heaven tonight?"

He just responded, "Honey, we need to go talk to the pastor before you do this."

When we arrived at the church, we went straight to the pastor's office. I sat down and explained to him what I thought the baptizing commitment was about. He and my father both began to weep. I didn't understand why they were crying, but they just couldn't believe that I had enough faith to give up everything I had on earth to go be with Jesus. My pastor explained everything to me and took me by the hand. He showed me the baptismal and the door that took you out the back once you were done. Then he kneeled down and gave me a huge hug. He told me how proud of me he was and that he knew I had amazingly strong faith in the Lord. Then he asked me curiously, "Were you scared when you thought you were going to die?"

I just looked at him and wholeheartedly answered, "No, I wasn't."

I had always loved the Lord. From about three years old, until the day I accepted Jesus, I remember telling all my friends, "Wishes aren't wishes, they're prayers to God." I also told them, "Every shooting star I see is God trying to get my attention." So, each time I saw a shooting star, I would say a prayer and ask the Lord for a mother.

My father wasn't dating much at that time. If he did go on a date, he would ask our approval and he always made sure they were Christian women. I was eight years old when my father took me to meet one particular woman. As soon as I met her, I knew that God had placed her into our lives for a special reason, and he was answering my prayers for a new mom. My father did marry her shortly after that, and I was overjoyed. I loved having a mother. Her joining our family was one of the happiest moments of my childhood. Unfortunately, the arrival of our new stepmother was not as much of a joy to my brothers as it was for me. They were teenagers, and after losing our mother, they weren't quite ready to endure this new change in their lives. They had many battles with her, and didn't accept her moving into our home as graciously as I had.

One day, when I was about twelve years old, my oldest brother came into my room, scooped me up in his arms, and gave me a huge hug. He looked at me and said, "You know, we don't tell each other we love each other enough." Then he talked about how we all used to go to church together. "Those were really good times," he said. I just smiled at him and agreed. Then he started talking about the Lord. He and my other brother had accepted Christ in their lives when they were younger. He told me that he knew where he was going and he wanted to recommit himself to Jesus. He was sure of his faith, and loved the Lord as much as I did. It was a great talk to have with my brother whom I adored so much. He said goodbye and left my room. About three hours later, my brother was killed in a motorcycle accident. That was the last conversation I had with him, and I was devastated. My brother was my best friend. We had always been so close, and I didn't know what to do now that he was gone. I was upset with the Lord that he had to die. Junior high school and puberty were already tough enough for me, but now I had to endure the pain and agony of losing another loved one.

I was heartbroken and confused, and didn't know how to handle all these emotions. It became a very questioning time in my life. My changing body, my overwhelming emotions, the pressure of school, and the unbearable pain in my heart was all just too much for me. I was doing poorly in school and was very depressed. I felt like the Lord had let me down. He took my mom, and now He had taken my brother from me too. I was struggling. I was drowning.

Soon, following this tragedy, the truth about my mother's death was told to me. I had always been led to believe that she died of a heart attack. Everyone must have thought I was now old enough to know what really happened. They shared with me her issues with alcohol and that her death was thought to be a suicide. Hurting and depressed, I felt so alone. I didn't have a clue how to handle all the pain that I was going through. This was too much for me to bear.

One morning, I grabbed a bunch of pills from the house: sleeping pills, aspirin and a variety of others. At the time, I was attending a private Christian school. As soon as I got there, I went into the girls' bathroom and swallowed all the pills I had put in my pocket. I didn't care who saw me. Like my mother, I saw no other way out. I had attempted suicide. I was only twelve years old.

I was rushed to the hospital and had my stomach pumped. After I was released, a youth pastor from the church started counseling me. We talked a lot about my suicide attempt and that it was not a way out. I even made a commitment to him that I wouldn't try it again. He told me I needed to deal with whatever issues came up in my life and to not run away from them. Although they were trying to help me, no one ever sat down with me and delved into what was really causing me this pain and depression. They just sort of said, "Don't do that again." I wasn't given any tools to tell me how to get through tough situations. It was just a quick fix to my problem, and it certainly didn't solve any of my underlying pain and misery.

Striving to make everyone like me, I had always been a people pleaser. After my ordeal with depression and out of control emotions, I felt I had disappointed everyone. Most of all, I had let the Lord down. I figured if I couldn't just give up like before, then I was going to be the best at everything I did. I was heading into a legalistic way of life, a totally opposite direction for me. I felt that if I went to all of the Bible studies, memorized the most Scripture verses, and was the best in the youth department then I would be able to get over these difficult feelings. I would definitely go to Heaven, and God would be very pleased with me. That's the basis of the theory I practiced during the rest of junior high and on into high school.

It was in high school that I started excelling in sports, and was chosen "Most Valuable Player" in each athletic activity. I still continued going overboard with my people-pleasing ways. The church youth group, along with the Christian high school, kept me very involved. Academically, I struggled. It became apparent during my junior year that I wouldn't graduate unless I attended summer school. Some rebellious feelings began to surface. I was full of anger and didn't know why. Instead of going to summer school, I transferred to a public school, figuring I would actually be ahead, rather than behind. Then, I would be able to graduate and wouldn't be a failure.

When I transferred schools, I lost a lot of athletic scholarships and couldn't play sports anymore. My main focus was now on church, my youth group activities in which I stayed very involved, and Bible studies. Not being able to go to my senior prom, I went with the youth group to Disneyland. While we were all hanging out there, the conversation naturally swung toward what we were going to do after graduation. Everyone was comparing their plans and goals, everyone, that is, except me. I just sat there feeling scared, confused and lost. I hadn't the faintest idea of what I was going to do next. Familiar, shaky emotions that I was still ill prepared to deal with built up inside of me again. I turned to my only refuge and prayed, "Lord, why is this happening to me? I'm graduating from high school, I'm still a virgin, and I've never done drugs. I just don't understand why I'm having these feelings of being lost and confused. Why is this happening to me?" It had to be my fault. If I had just listened long enough, or focused on Scripture deep enough, I would have heard what the Lord wanted me to do. I had been good, and I wanted Him to fix it all for me right then and there. I wanted the confusion to just go away.

It was then that I started acting out my addictive behaviors. I was overeating and then taking diet pills uncontrollably. Everyone in high school was doing it. I was just fitting in.

Graduation arrived and I made it! To celebrate, our family went to Baskin Robbins for ice cream. I remember feeling rather proud of myself. I had chosen to be with my family instead of with many of my friends and fellow graduates who were going out and getting drunk. That is, until my dad asked me that age-old question, "What are you going to do now?"

I just looked at him and said, "Well, what should I do? What are the rules to life?"

He just said, "Well, hon, there are no rules to life. You learn as you go. You try new things, and you always try to do what the Lord wants you to do." All I heard was "there are no rules to life."

I asked him, "Do you think I should go to college or work? What exactly do you think I should do?"

He answered by asking me the question, "What do you want to do?"

I didn't know. I felt an overwhelming fear come over me like I'd never felt before. Again, I felt alone, out in the world by myself, and angry with the Lord. My heart was starting to harden.

It was during that summer that I started disconnecting myself from God. I helped out at a summer youth camp, but it was with an insincere heart. I didn't memorize my Bible verses and I sure didn't have much excitement for the Lord. At home, I quit going to church and quit reading my Bible, which once were my favorite pastimes. I was still taking a lot of diet pills. The only constant was the feeling of being completely lost.

About three weeks before fall quarter started, a friend of mine from the youth group called and said the volleyball coach at the local college wanted me to try out for the team. I made the team, and enrolled for classes to be eligible to play volleyball. That was now my main focus and only goal, over excelling in sports again. Thoughts about serving the Lord were nowhere on my radar screen anymore.

It was after a volleyball game that I went to my first college party. I walked around and mingled, but didn't drink. I remember watching those who were getting drunk, thinking to myself, "Man, they're having a good time. What's that all about? I want to feel like that instead of the way I've been feeling." Two weeks later, I was invited to another party. That one night changed my life forever.

I had my first drink, while also trying about four different kinds of drugs. As soon as those drugs entered my system, I was hooked. There was no stopping me. The euphoria they induced couldn't be beat! For nineteen years I had lived a good, moral, Christian life. At that one party, I had my first taste of alcohol and drugs, and was transformed into an addict then and there. That night began a cycle of addiction that would haunt me for the rest of my life.

In the first stages of my drug use, I used cocaine and methamphetamines. Quickly, my drug of choice became crack cocaine. I sought to always be high, and so, just as quickly, I was unable to make it to my classes. Consequently, my college career only lasted about a year and a half.

With college out of the picture, I found work at Costco while remaining a hard-core crack addict. About this time, I tried going to church a few times, but I just couldn't do it. I was filled with so much guilt and shame, I couldn't talk to God or even think about Him. I finally quit going to church with my family, even avoiding family Scripture readings.

My crack addiction continued to get worse. Even though I was still living with my parents, I soon began bringing my drugs home with me. By this time, I had put God in the back of my mind, high up on a shelf. At one point, I took my Bible and shoved it in a drawer where it couldn't haunt me. That was my final farewell to the

Lord for many years. Each morning, instead of filling my mind with prayer and thoughts of thanksgiving to God, I focused on my addiction. Where was I going to get my drugs today? Where and when was the next party? How am I going to make it through work today, while high on drugs, and still be able to make enough money to support my habit? As time progressed, these were the only thoughts that occupied my mind.

The way I was living by now was extremely gloomy and agonizing. An addict's life can be summed up in only one way: a constant, living nightmare. I witnessed many horrifying things in those few years, but the baby being born in a dope house bathtub was probably one of the most upsetting. Predictably, considering how heavily I used drugs during those days, the day came when I took too large of a dose of crack and ended up overdosing.

I was in a dope house at the time. When my overdose was noticed, my fellow junkies filled the bathtub with ice water, and then put me in it, leaving me there until I was somewhat revived. Then they got me out and laid me on the floor, covering my entire body with a towel. I couldn't move. The towel was covering my face and I couldn't see, but I could hear them talking as they stood above me. They were discussing what they were going to do with my body. They expected me to die, and then they were going to just dump my body somewhere else. They rifled through my belongings, took what they wanted, and then just left me there to see if I'd come out of it. After several hours though, I was able to get myself up and drive home. As horrible as that experience was for me, I still hadn't hit rock bottom. I continued living my waking nightmare.

Another vivid experience occurred some time later when some members from a local gang, the Crypts, asked to borrow my car. They said they'd give me some dope in exchange for using it. I told them no since I was still on my dad's insurance, and I didn't want him to find out about my lifestyle. The lead gang member pulled out his gun, pressed it firmly against the side of my head, and said, "What choice do you have?"

I immediately gave him my keys and said, "Here, take the car." They drove off and left me at the dope house. I stayed there for three days and was high the entire time. I waited for them to actually bring back my car, naively thinking they would be kind enough to do that. When the drugs at the house ran out, I had to call my dad to come get me and take me home. I didn't want him to see where I was, so I walked two or three miles down the road for him to pick me up. My dad called the police, and eventually the gang members brought the car back. I didn't press any charges, or make a big deal out of it, knowing it would likely cause the gang to threaten my life or the lives of my family.

My parents didn't have a clue about my drug use the entire time that I was strung out on crack. Since I was now old enough to drink, my dad started to wonder if I might have a drinking problem. I felt so ashamed to look at my parents and spend time with them, that I began avoiding them completely. They were still very warm and open to me, but I totally closed myself off from them. I would get up really early so I could leave before they got up, or I would stay in bed late, waiting until they left the house. I was doing everything I could to hide my drug use from them. I was living a double life in so many ways; it was really beginning to take its toll on me.

It took until I was about twenty-three before my family finally caught me. My father and stepmother were enjoying a typical Sunday by reading Scripture in the living room. Around two o'clock in the afternoon, my brother and his wife came over. She worked in the area of psychology and had known for a while that I had a problem, unlike the rest of my family, who were in denial. Casually, she asked my dad where I was, and he replied, "She's still in her room, sleeping."

"She's not sleeping," was her reply before she paused and said, "I'm so sorry I have to do this. I'm sick of this." Moving over to my door, she pounded on it and yelled, "Come out here, now!"

I yelled back, "Go away, I'm sleeping."

"I know you're not sleeping in there!" she called back. My dad stood up, telling her to leave me alone. She just looked at him and kicked my door open. Grabbing my father by his face, she forced him to really take a long, hard look at me for the first time in a long time. There I was, sitting on my bed holding my crack pipe, with all my drug paraphernalia lying around me. I was five-foot-eight, no more than one hundred twelve pounds, and a ghastly sight. There was no hiding now. I was totally exposed.

"Now you can see what she's doing," she said to my father. "You can't be in denial anymore." I was so utterly ashamed. It was by far the most painful time for me, knowing that my dad had finally seen the truth and depth of my problem. It was just as awful for him. He had gone through all this before, when my mother was depressed and on drugs. The outcome with my mother had been horrific. He knew the pain she had suffered with psychiatric treatments and shock therapy. He didn't want to admit or even think about the fact that I was going through the same thing she had battled over twenty years ago. My family gave me two options: either go to treatment or go to jail. I definitely did not want to go to jail.

My family chose the rehabilitation center for me. I went into it reluctantly, but comforted myself by thinking, "I'll just go for a while and then go back to 'using' again as soon as I get out." It wasn't a drug therapy center, but a psychiatric facility. It was an absolutely awful time for me. For the first time in four years, I didn't have any hard drugs in my system, and I was uncontrollable. I had to be locked up, placed in a padded cell, and was even put into a strait jacket a few times.

Eventually, I was released from the psychiatric clinic and forced to attend Narcotics Anonymous, using its 12-Step Program based on the Bible. At that point, just hearing the words "God" or "Jesus" made me uncomfortable. I didn't want to hear anything about religion and grace. I was still too full of shame and guilt to think about the Lord. Knowing how poorly I had lived my life up until then, and having once been a dedicated Christian, made it that much more difficult to come back to the Lord. Still, the 12-Step Program began to sink in a little. It was really beginning to help me with my addiction, so returning to the Lord was becoming easier to consider. I was actually able to say the word "God" after many years of avoiding it.

Shortly after that, a friend asked me what my "higher power" was. I just looked down and mumbled under my breath, "Christ."

"What?" she asked.

"It's Christ," I said, a little louder.

"Oh! Me too!" she cried.

I didn't want to be too loud; I didn't want anyone to know I was a Christian. I was a horrible person who didn't deserve to be called a "Christian."

I was clean for ninety days before I finally felt ready to go back to my old church. I shared with members of the congregation what I had gone through, and what was going on in my life. Some of them were really supportive of me, proud that I had gone to treatment to get clean. Others did just as I had thought they would, they shunned me. They were disgusted that my life had gone in that direction, so they offered me no support. The few that did offer their support told me to just keep coming to church. They didn't feel I needed the 12-Step Program. They said they would take me under their wings, and their support would see me through. All I had to do was accept their help, and keep coming to church. So, I did just that. I quit going to the 12-Step meetings and quit connecting with others who were also going through recovery.

Three short months later, I relapsed by hitting my pipe once again. That woke me up to my need to continue with the 12-Step Program as a part of Narcotics Anonymous. This time I stayed clean for nine months before using drugs again.

This wasn't working, and having experienced the rejection from some of the people, I told myself I would go back to church only when I was clean and had wholeheartedly given my life back to Jesus. Having always been an "all or nothing" person, being on the fence, so to speak, didn't suit me at all. I continued with the 12-Step Program to help me get clean, but it was a lot harder to figure out how to reestablish a relationship with Jesus again, and it took me several more years to decide where I stood with the Lord.

In the meantime, my mother's brother and his wife offered me a job, and urged me to move to Washington State to live with them. My uncle had witnessed what had happened to his sister, and did not want me to end up the same way. He felt a change would help. Seeing a chance at a fresh start, I left California.

About two months after my arrival in Washington, my brother experienced some really heavy depression, and ended up separating from his wife. He then committed suicide. To be hammered again with the death of a loved one, with only nine months recovery to my credit, laid me out flat. I had been working so hard and was beginning to do well, but still bad things continued to happen to me. Resentment for the Lord welled up inside me, and I blamed Him. Losing another brother hurt tremendously, but what was most painful was realizing he had followed in my mother's footsteps. He gave up, leaving behind three young children, who now had to go through life without their father. By age twenty-six, I had lost my mother and both of my brothers. My brother's suicide devastated me. I sought help from a psychiatrist, but the hurt ran too deeply and I relapsed for the third time.

Needing to escape family, I moved to Cle Elum, where my uncle had a cabin, and attended massage school. Soon after, I met my future husband and we began dating. I was still "using" very heavily, and all my money was spent buying drugs to support my habit. Since I needed help paying the rent, I was living in an apartment with two friends of my boyfriend. Supporting my daily habit had put me on the verge of losing my job. I was in a very dark place, both spiritually and emotionally.

One day, as I was about to go to my dealer for more drugs, something told me I would probably end up sticking a needle in my arm, overdosing and dying. Just before leaving my bedroom, I looked up and prayed, "Lord, I don't want to die. I don't want to die out there. Please, please help me. I don't know what to do." Without conscious effort, I walked through the door. But instead of leaving, I went to the two men who shared the apartment with me and began telling them about my addiction. I told them I needed help. Finally, I sat down in front of them, bleak, sad and empty. I must have been quite a sight with my pasty, white skin, eyes bulging out of my bony skull and my fingertips and lips dry and burned from my daily crack cocaine use. All in all, I'm sure I looked like a walking corpse.

"We thought there might be something wrong," one of them said.

I don't know if these men were Christians or not, but I know God placed them in my life to finally help me get through my addiction.

They asked me, "What do you want us to do? How can we help?"

I responded, "I need to go to a meeting where there are other recovering addicts, and I need to get right with God."

They just said, "Okay," as they shuffled me out the door, put me in their car, and found a place where a Narcotics Anonymous meeting was being held. For the first time, I spilled my guts, finally sharing my feelings from my heart. God completely rescued me that day.

God was saying to me, "Let Me help you. You are honestly asking for help from the depths of your heart." It was a spiritual awakening for me. I firmly believe that accepting God's guidance to appeal for help from those two men saved my life. That was the day my life finally began turning around.

The next two weeks were extremely grueling as I went through detoxification by myself, while still going to work every day. Most addicts detoxify in a clinic under support and care from professionals. I did it physically alone in my apartment, but with Jesus by my side, and He made all the difference. I went through horrific things that most people couldn't even imagine. I hallucinated, thinking there were things crawling all over the walls. I heard voices speaking to me that weren't there. I vomited, shook continuously, had terrible headaches and my bones ached. I was unable to sleep for days on end. Not only was it physically painful, but also emotionally draining.

Soon after that, my life began to get back on track. I was still with my boyfriend. By the grace of God, he had not abandoned me during this horrible time. We began living together and about three years later, I got pregnant. I had suffered a miscarriage years before when I was still "using." Because of my drug use, the baby had been deformed. It was difficult for me back then, but I feel so blessed now that I never actually delivered a baby during those awful years. It would have been the one thing that would have permanently kept me away from the Lord. I felt so blessed by God to be given this chance to have a child, now that I was clean. When I was a little girl, I had dreamed of being a mother. Now, I was going to experience the joy and happiness of motherhood without the involvement of drugs.

I knew that living with my boyfriend before we were married was a sin, and now here I was pregnant. I thought there was no way I could go to church, especially now. It brought up the old feelings of guilt and unworthiness. I constantly condemned myself for all these things, inadvertently pushing myself that much further away from the Lord. I read Scripture and prayed to the Lord at home, but I wasn't sure if I was ready for church yet. We were living across the street from a church in South Cle Elum, and every Sunday I could hear their beautiful music. It was beckoning to me, calling me to come over. I'd sit by the window listening, my heart aching, wanting to cross the street. Finally, one Sunday, I got the courage to go over to the service. I planned to just sneak in, sit in the back, and hurry out when it was over. I managed to get in without being noticed, but it wasn't so easy sneaking out. Numerous people stopped me to welcome me to the church. All I could do was hope they only saw my face and not my bulging belly, especially with no ring on my finger. Even the pastor came over and greeted me that first day I attended church.

I went again the next week, and after the service, I thought seriously about attending there regularly. I still felt very ashamed, and was afraid that I would be judged if I continued going as an unmarried woman who was pregnant, not to mention being a recovering addict. One Tuesday afternoon, while sitting in my bedroom, I was overwhelmed with a tremendous need to go to a Bible study. I prayed, "I want to recommit myself to You, Lord. I want to get involved in the fellowship. I want my child to grow up knowing You. I want to get married, and I want to get married in a church."

So I walked across the street to the church, peeked in, and asked if the pastor was available. A lady led me to his office, and I slowly walked in and stood in the middle of his room. I put my hands on my hips and said, "I have some questions for you. I want to go to your church. Are you going to kick me out if I go to your church?" I paused. Continuing on, I said, "I'm a recovering drug addict, I live with my boyfriend, I'm pregnant and we're not married. Can I go to your church?" I waited a little, expecting him to just kick me out right then and there.

He just kept the same look on his face and said, "Why sure you can come to our church. You don't want to teach Sunday school or anything do you?"

"No, not at all," I said.

"Well, that's fine. You come right on over. We'd love to have you." Then he got up and gave me a hug.

One year later, the pastor married my husband and me. At the same time, we had our baby's dedication service. After I'd been clean for about six years, we tried again for another baby. Unfortunately, I ended up having numerous ectopic pregnancies, a few surgeries and nine miscarriages. It seemed as if God had given us that one miracle child and that was the only one we would have. As a last resort, we tried invitro fertilization, hoping I would get pregnant and be able to carry it full-term. God displayed His great grace in my life. The day I celebrated ten years of being clean, I found out I was pregnant with twins! I was overjoyed and grateful that the Lord had continued to bless my life.

Our twins are two years old now, and in June of 2006, it will be my thirteenth year of being clean. It has been a long haul for me, but now that the Lord is by my side, I am able to make it through each day without drugs. A doctor once told me if someone is a true addict, they have to have the physical compulsions, the mental obsession and spiritual malady. To conquer the physical compulsion, I quit "using" and spent my time with other people who weren't drug addicts. I got through the mental obsession by reading the 12-Step material. After I addressed the first two issues, I was able to reconnect with the Lord, and shed that spiritual depression that had haunted me. Jesus brought me back and saved me. I still struggle a little with the mental obsession. I know that if I let myself, it will overtake me in all areas of life, not just around the issue of drug use.

The Lord is my addiction now. I go to Him every morning, sit down with Him and pray, "How do I stay clean today? What do You want me to do for You today?" If I don't connect with Him each morning, my day is a mess by three o'clock in the afternoon. I have to stop and say, "What is going on? Oh, I know, Lord, I didn't talk with You today." Then I take the time to pray and speak with Him.

I still go to 12-Step meetings and conferences. I don't want to go back to the lifestyle that I once led, and I know that I will stray without the Lord right next to me. It's something that will haunt me forever. My only cure is God. He is my salvation in every sense of the word. He is what I can't live without. I feel blessed that the Lord has helped me to escape the cycle of drugs, which once controlled my life. He dominates my life now, and I know I get through each day with Him by my side.

5
God Bless America
Written by Marlene Drew

My father drew the curtains so tightly that light could neither escape nor creep in to reveal what was taking place in our home. A tiny thread of anticipation wove its way up my spine. I knew full well the danger, but was unable to resist the thrill of treading into forbidden territory. I was young, but aware of how serious the recriminations would be if we were caught in our secret. I vividly remember that day in the year 1981, for two reasons. The first was the launching of the Columbia Shuttle that we all watched with great interest. The second, but most important to me, was that for the very first time, my father and I were going to listen to the "Voice of America." In my country, we loved to hear what was going on in the United States, the country that shone brightly as a beacon of hope to those of us living in a world that was not free. All these years have gone by, but I still recall how it awakened a sense of excitement in me.

We were in our living room, the same one I had always known, where dark, mahogany cabinets stood. A thick, deep-red carpet covered the floor, and a soft-cushioned couch sat in the corner. A dining room with a mahogany table and six padded chairs was off to the right. I could see into the kitchen through the glass wall that divided the room. My father and I were seated in chairs with a little table in front of us, where exquisite

Bohemian cut crystal glasses were placed. One half-full glass held Martini Bianco, on the rocks, with a twist of lime. The other held just a splash of the same drink. Sitting in the center of the dining table was a small battery-powered radio. This was home. Everything looked the same, but tonight there was something electric in the air. I was going to get my first unadulterated earful about the truth of freedom in the United States of America.

My world growing up was colored in shades of gray. There were no bright, vibrant rainbow hues symbolizing joy, laughter, or mirth. Neither were there the dark tones for contrast, to represent the sad or hard times. Nothing existed but gray and more gray, like living under a storm cloud that brightens a bit, or darkens with more storms on the way. It was similar to the fog that occasionally settles during an inversion in our valley where I now live. It was a time when you thought you would go mad if you didn't see some sky, a color, or anything to break the monotonous, never-ending gray. I am not sure whom it was hardest on, the older generation who still held the memories in their hearts of a free nation, or the younger people to whom freedom was only a voice on the radio. Stories had to be told in secret, in hushed tones, amongst trusted family members. All of this gray has a name: oppression. The dictionary defines oppression as "the exercise of authority or power in a cruel or unjust manner." But in order to see it, you need only look at the countenance of the masses and see their closed, guarded, unsmiling faces. You could also observe their interactions with one another, always keeping emotions in check, and eyeing each other warily, exchanging furtive glances that breed mistrust like a fungus. One was always wondering who might turn you in to the authorities under some misguided premise of great reward. Suspicion ran rampant. Which neighbor may have turned communist? It was the unknown that drove these wedges between friends, neighbors and even family members.

Czechoslovakia was an enterprising, self-sufficient nation before the Russians were awarded control. My country only enjoyed true freedom between the years 1913 and 1938. We were a democracy, proficient in manufacturing and agriculture. There was an intelligent management of the farming communities, which provided enough food for everyone, and nothing needed to be imported. We produced cars, firearms, machinery, shoes, jewelry, crystal and more.

At the close of World War II, when Germany was defeated, the allied powers, Roosevelt, Churchill and Stalin, the great triumvirate, met to divvy up the bounty, so to speak. Churchill was outnumbered in his desire to keep our country in the European Bloc, so it was determined that Czechoslovakia would be under the jurisdiction of the Eastern Bloc countries. Stalin was in power in Russia, and he became the governing ruler over the Czech Nationals. We had a "puppet government" in place (what we were called on the red carpet in the Kremlin) to make it appear as if we had some power to make decisions for our own country. But in truth, we had no say in our own destiny, and thus became a Communist nation. We were reduced to a "Banana Republic" economy and became a satellite, or slave, to the Kremlin, a country that couldn't hold a candle to us as far as technology or anything else was concerned.

My parents were highly educated people. My father was a forestry engineer and my mother was a physician. Our family, as a whole, was educated, intelligent and observant. We were not rich or considered well off, even though my parents had high-profile jobs. Each profession had a set wage determined by the government. If you do this job, you get this pay, and if you do that job, that is your pay. It was all standardized according to the government's idea of your worth. They determined your professional and personal value. This kind of oppression promotes mediocrity and suppresses excellence. It sets the standards of what you can do and what you will be allowed to achieve. It keeps people in a box.

Everyone had *the bread and the soup*; no one was going hungry. We had jobs and were not starving, at least not in the physical sense. We literally lived behind an electric fence, a barrier that could not be penetrated. Our country was occupied in every respect. We could travel freely inside Czechoslovakia, but had no freedom to roam outside the fence. We were not a happy nation. We did not struggle physically or economically, but suffered through ideological indoctrinations and a moralistic system. It was very insulting to our personal freedoms. In high school, we would reach the end of a long day and be ready to go home. Occasionally, the teacher would come in right at the last bell, and tell us that we needed to gather in the main square with all the other schools. Of course, we would ask why. We would be told, "It is none of your business; all the schools are going to gather. It is mandatory; no one is going home." When we would all amass, there would be a staged demonstration and a speech given by some idiot from the government who would be ranting about the "American Imperialism" expanding in Europe. A photo op would be made, making it look like we were all in agreement with our government. It was total manipulation, very insulting, and an all too common occurrence.

We had our own version of the Secret Police, similar to the KGB. They were always on the lookout for someone expressing political opinions or ideas, or acting in any way that might be deemed anti-government. Being turned in to the authorities was always a possibility. There was the mistaken premise that somehow the mediocre could rise in the ranks simply by turning in a dissident neighbor. The sad fact was there were no winners. Everyone lost, living under oppression.

Ice hockey, being a huge winter sport in Czechoslovakia, was one of the few ways we could ever beat the Russians. If we did win, it usually started a small riot, and created a minor uprising in the country. It gave us a small feeling of victory. One time, an ice hockey match was being held somewhere in West Germany. The Czechs were there, as well as the Russians. In the stands, some previous defectors from my country held up a huge banner that translated to "Beat the Russians for all Eternity." This was a paraphrase of the banners seen daily in our hometown, which translated to "Friendship (Be) with the Russians Forever." It was essentially a parody, "Beat" instead of "Be with." A good friend of mine, who was playing on our team, found a drop cloth near the locker room that some painters had been using. He took a black marker and wrote the slogan "Beat the Russians." The painters turned him in. As a result, they got nothing, but my friend's future was radically changed. The Russians severely limited his choices for the future by allowing him to go to a mining school, certainly not his choice of careers. I must emphasize, there was nothing on this drop cloth that a grandma or an eight-year-old child could not read. It contained no expletives, nothing profane, and no obscenities; it was strictly a polite, political expression. I did not have any single defining moment in deciding to defect; it was a gradual buildup. I knew that this was no place for me, and this incident was just one of many that culminated in my decision to leave.

The only people you could call somewhat free in our society were the very young and the elderly, especially the elderly. What can you really do to a person who is eighty years old? It became their job to carry the truth and tell what it was like to live in a free world. The government suppressed them as much as possible, but could not stop them entirely. I was lucky enough to have a set of grandparents that took to heart the responsibility they felt, which was to tell me the truth. My grandmother also took me to Sunday church and talked to me about religion. It was very important to her that I learn about God and Jesus. She shared with me the little bits of Christian literature that she had saved from her previous days of freedom. These were little Catechism books and other children's books about God. Her religious teachings really stuck with me. My grandfather was more involved in educating me about the personal freedoms that no longer existed. I spent most of my formative years, ages two to six, with my grandparents since the government required that all mothers had to go to work. The firm foundation and belief in God that came from those years with my grandmother never left me.

Going to church on Sunday was permitted, but you could be sure if you stepped into one, your name was on a list. There was a kind of street organization that kept tabs for the government on who attended which church. About seventy to eighty percent of the population was Roman Catholic, followed by Russian Orthodox and Protestant. Every move was monitored to make sure there were no extracurricular meetings, or any attempt to organize one, that stepped outside the allowed religious activities.

Despite all the anti-American propaganda that bombarded us, as often as we could, our family would listen to the "Voice of America." This was always extremely risky. Children could be taken from their homes if the government didn't agree they were being raised for the "common good of the nation." Hearing that phrase still makes me shudder. The public school system would try its best to indoctrinate and influence us, but in spite of this, my parents and grandparents instilled in us an unshakable ideology about freedom and what it was worth. So, the more the government tried to push their propaganda down my throat, the more determined I became to leave my country and try to get to the United States. By my third year of high school, I knew that I was leaving Czechoslovakia, and I knew I would be going alone.

My mind was set and I started to formulate my plan, but I also realized it would not be easy. I decided that I would try to make it to America, Canada or Australia, in that order. When I talked this over with my parents, they were, for the most part, accepting of my decision to defect. My mother really wanted me to stay and finish college first, but she didn't try to change my mind. She wanted to be sure that I had a clear picture of what life was like out in the big world, and that I did not have some idealistic view that the streets were paved with gold. She wanted me to know I would have to work hard for a living.

After completing high school, I was lucky to be chosen to go on a trip with a youth travel agency. It is somewhat similar to the senior trip that kids in this country take. We would be traveling on a bus to Yugoslavia to the shoreline of the Adriatic Sea, near Dubrovnik. It took me a full year to get all the necessary stamps, permits and paperwork ready. I had to get recommendations from my school officials stating, "Yes, he is an okay guy; he can be trusted." Approval was also necessary from the National Youth Organization (the Communist equivalent of the Hitler Youth Group), as well as other similar groups. We would be going as a well-chaperoned group of kids to spend some time at the shore. Once there, we would stay in a bungalow for a few weeks and then everyone would go home. Well, almost everyone. I had no intention of returning.

The night before we left, I went into one of the cathedrals in my town and prayed. I asked God to watch over me, give me guidance, and to bless my family that I would be leaving behind. When morning came, I said my goodbyes and boarded the bus to take me to what I hoped would be a new life, a life of freedom and opportunities.

The Adriatic Sea is a beautiful arm of the Mediterranean Sea, with sparkling, azure blue water and a bold, rocky shoreline. It is dotted with big, light-gray boulders. It separates the Italian Peninsula from the Balkan Peninsula, and the Apennine Mountains from the Dinaric Alps. These mountain ranges provide the backdrop to the whole amazing panorama.

I spent a couple of days at the shore acting like any other tourist, but I was really gathering information, as well as my mental and emotional energy. I knew that once I made my break, there was no turning back. My resolve was strong. I looked around the area and got the lay of the land. I knew where the bus station was, what road went where, and everything else necessary to be successful in this attempt to escape out of the country. When I felt ready to go, I told the travel guide that I was going to hitchhike around the countryside for a couple of days and then return to the group. She wasn't too keen on the idea, but didn't have any choice. I was no longer in Czechoslovakia, and she really couldn't stop me. Yugoslavia was the freest of the Eastern Bloc countries at that time, and I could travel around as I pleased.

After figuring out the bus schedules, I caught the first Greyhound-type bus heading straight North to the Italian border. Traveling the shoreline highway was beautiful with the sun setting over the water and reflecting brilliant colors off the sea. It gave me the opportunity to reflect about my own life and the decision I had made to leave. Nothing was going to deter me from that goal. I was on a journey to embrace something that, until now, had been no more than a voice on a radio or a whispered secret from some older person who knew what freedom felt like. There was a keen sense of anticipation to be getting on with my business, but there was also a sense of sadness, not knowing when, or if, I would ever see my family again. I was embarking upon a journey from which I would not return. I had a feeling that I was losing my known universe. It hit me with a jolt. This was my nineteenth birthday, and I could live a very long life with no contact from my family or my good friends and acquaintances. Never again would I be able to see my home or any of the parts of my country that I loved.

The village of Koper, Yugosolvia, was the last stop. As I exited the bus, there were many feelings rolling around inside me that I am unable to describe to you. How do you explain spaghetti sauce to someone who has never tasted Italian cuisine? Indescribable is the only word that fits here. It is a totally different state of mind.

I was still a long way from the Italian border, about fifteen miles, and I would have to walk all night and part of the next morning to get there. I didn't know if the border was fortified or not, or what I might find when

I got there. I have tried to think of words to best describe my state of mind that night, but I have come up with nothing that seems adequate. It was almost surreal. It was like looking at a three dimensional picture. First, it seems very flat until you scrunch your eyes just so, and suddenly you find yourself inside of it, and a whole new dimension is opened up to you. There I was, inside a picture of something that I had only imagined for years. Now I was trying to get the feelings inside of me to match the vision. I prayed every step of the way, asking God to look after me and be my guide. More than once, I prayed a prayer of thanks for my grandmother, and for her role in teaching me to trust in the Lord. Without Him, I don't think I would have ever found the courage to take such a step. He gave me peace about my decision to leave, and gave me the strength I needed to carry it out. I was thrilled when I found that the border was open. I was able to walk right across all the way to the city of Trieste in Italy.

Once there, I found the train station to get passage to Padova, where I turned myself over to the Italian police and asked for protection. It was required that I go through a government agency of some sort in order to be granted political asylum. There were two men behind the desk who didn't understand a word of what I was trying to say to them. It was obvious that there was more between us than a wooden barrier. I was asking for asylum in my language, and they were asking me what I wanted in Italian, a language I couldn't speak. After a lot of gesturing back and forth, I wrote a note on my notepad and handed it to them along with my passport. That seemed to do the trick. They took me into the inner office and started to process me, as if I had been arrested, height, weight, brown hair, brown eyes, that sort of thing. After the paperwork was completed, they escorted me back to the train station. I was put on a train to the city of Latina, where a camp for Eastern Bloc refugees was located.

The camp was an old Army barracks, holding about twelve hundred people who were all waiting for their fate to be determined. Our movements were not restricted, even though there was a gated entry and a fence all around the yard. The nationalities predominantly represented were Polish, Romanian, Czechoslovakian, Albanian and a couple of Bulgarians. The camp was a "Babel" of languages, cultures and attitudes, and our nationalities didn't blend a whole lot. Because of the language barrier and our different customs, each nation, more or less, kept to its own group. I didn't have a prayer of understanding the Romanians or the Albanians. Every individual in there looked nervous and seemed a little impatient. If any animosity existed between the countries, I didn't feel it. I think we all had the same thing on our minds, freedom. We just thought about it in many different languages.

When I arrived in the camp, an organization called the IRC took me under its wings. Its function was to walk the refugees through the necessary paperwork and set up the interviews. Since the IRC knew the ropes, they explained what was going on, and what had to be done to get through the mountain of paperwork and forms that had to be filed. They had offices in each refugee camp, with their main headquarters in New York City.

I spent about six months in the camp, in perpetual limbo, waiting to be called for one of many interviews to check out my story to be sure I wasn't a spy or some kind of deviant. I could only wait since I didn't have the necessary papers to work. Eventually, I was asked to state which country I wanted to go to, giving me the choices of Canada, Australia or the United States. I filled out the paperwork with the United States as my first choice, then Canada and, lastly, Australia. These papers were then sent to the Embassy of my country of choice, and that country processed the paperwork. I had no knowledge of what happened after that, but I was eventually called for the long-awaited interview.

My next step was to travel to the U.S. Embassy in Rome to meet with, and be questioned by officials about my reasons for defecting. They wanted to know about the oppression I had lived under, and my political reasons for wanting to leave. They needed to be convinced it wasn't just because I wanted a new car or something similar. I had entered the camp in June of 1985, interviewed at the United States Embassy in Rome in November, and received their decision of acceptance to the United States in December. I couldn't believe it had finally happened. I had been living in this "no man's land," a literal man without a country, for six long months. I had put my faith in the Lord and He rewarded my trust.

After my acceptance to the United States, the IRC

took over, making all the arrangements for me. It was a common practice to scatter the immigrants to different places, unless they had private sponsors (which I didn't). I was provided with an airline ticket to San Diego, California. The IRC Regional Office moved a group of us into an apartment, paid the rent for the first three months, and took us to the employment agency to help us find work. My first job was as an apprentice refrigeration mechanic.

I stayed in San Diego about six months, but never unpacked while I was there. A powerful emotion began to settle in, a feeling that I was no longer within driving distance of home. I was almost in a state of shock. I had finally made it to the United States, but I still felt as though I were in survival mode. As soon as they would drop me off at the apartment each day, I felt like I was on the run. I wanted to get on a bus and go somewhere, anywhere, and didn't know why. It was a very irrational and persistent feeling. It wasn't homesickness that kept my emotions in a state of unrest, but the fact that I had been so long without a place to call home. I couldn't shut off the feeling that I was still waiting for something.

My need to keep moving landed me in Hawaii for about six months. While there, I met some people who fished in Alaska and they offered me a job as a commercial fisherman. I spent that summer up north, gill net fishing, and from there, moved to Washington, near Seattle. While I was in Italy in the refugee camp, I studied maps of the United States and researched the geography of the different regions. The Northwest was the area that had interested me the most, and was the place where I had chosen to live and to build a new life. It was far enough away from Europe, and had the best climate, people and opportunities.

During this time, I hadn't found a church I wanted to attend, nor had I become part of any religious organization. I had always gone to the Roman Catholic Church in Czechoslovakia, so naturally, I tried that denomination here in the United States. Their beliefs and practices were very different from how we worshipped in Czechoslovakia. When a friend of mine took me to a Christian church in Marysville, I found it was more in line with what I was taught as a child about the Holy Trinity. It was at the church in Marysville where I met the woman who would later become my wife.

While living in the Seattle area, I realized that the environment was not what I thought it would be. There were only two seasons, wet and dry, and I missed the other two. Eventually, we moved to Cle Elum, Washington, and I must say, this place is everything I hoped and dreamed it would be and more. We now have two beautiful children who enjoy the freedom of this country. I pray that my family will never experience living under communistic rule, as I did in Czechoslovakia.

If I were to list for you some of the most impacting moments of my life to date, then these thoughts would definitely be at the top:

On that long night when I was walking into Italy, there was no definite line to mark the border. I was praying the entire night that God would protect and guide me. When I finally looked out across the bay and saw the lights of Trieste, I knew He had answered my prayers. I was in Italy and I was safe.

In 1992, I received my US citizenship. I had prayed for this moment since my third year of high school. I had put my trust in God, and here I was, a citizen of this beautiful, free country. It was the forbidden world that, by all rights, coming from Czechoslovakia, I should never have been able to set foot upon. This was a "dream come true!"

I wish everyone living in America would realize how important he or she is to the rest of the world. Do not buy into the propaganda that you are corrupt and bad. There will always be a bad element, wherever you go. You have no idea how the country of the United States of America is a shining light to the rest of the world!

May God bless each and every one of you, and may God bless America, the land of the free and the home of the brave!

6

Just to be Loved Unconditionally

The Story of Miriam Greenman
Written by Tina Hansberry

I pulled myself out of bed, gathering all the strength within me to begin a new day. The sun was beginning to peek out from the Oregon skyline, and it appeared that I actually had a shot at having a "normal" morning. I went through my regular routine of getting ready for work and organizing my things. Saying goodbye to my roommates, I grabbed my bicycle and left for work. How was I to know that upon returning home, my whole world would begin to unravel?

As I approached the entrance to my home, I noticed the door was ajar and I could hear voices inside. I slowly pushed it open to find a detective questioning my roommates. They all looked at me as I entered the room. The detective introduced himself and explained what he was doing, and that he had been waiting for me to arrive. I noticed the strained smile upon each of my roommate's face, and sensed this huge knot in the pit of my stomach. I felt like I was suffocating. Without even knowing it, I was about to be the suspect in a long line of crimes. Worse, the "real" offender would soon be revealed as the person responsible for the baby I was unknowingly carrying. This unborn baby was the product of betrayal, a date rape baby by definition, which brought guilt, laden with my own doubts. My story really began much earlier when glimpses of possible hope still existed in my life. This day, however, was the start of a downward spiral of complete despair. I was oblivious to the hell that was to come.

My childhood was not unlike any of the other children on my block. I was raised in a disciplined and controlled Christian family, with a pastor for a father and a nurturing mother. Yet a void remained. The relationship with my father was strained. The caring trust had been broken years earlier, and I was left with little more than disdain. He was going through his own dark season when my teens hit with the fury of emotions that only a teenage girl can relate to. The gulf broadened. I had desperately sought his approval over the years, only to come up short. Thus, the conflict-filled path of life began.

All of the securities of my life began to be pulled away from me. At the age of eighteen, things went from being consistent and predictable to being fearful and confusing. My first acceptable boyfriend broke my heart days before I was turned loose on my own. I wasn't ready! My life was so governed and structured that I wasn't sure how I would make it. I only knew I had to try.

A few days later, those thoughts returned when I saw my parents pulling away in a U-Haul truck to begin their new life in Seattle without me. I had lost my job that day and didn't have the heart to tell them. Arrangements for my new life had been set in place, and they offered me the support they could. It was hard for them to let go of their last child, but, with faith, they hoped it would be all right. The sight of them leaving created fear of the unknown, and my predisposition to depression began to rear its violent and life-consuming head.

Life rolled on down the highway, as they say, and I lived from day to day, just trying to get by as the adult I was expected to be. My best friend in the entire world would soon be next on my list of losses. We spent great amounts of time together throughout high school. She was always in trouble here and there, and needed someone to look out for her. "You are my voice of reason," she would sometimes chide me.

We shared many adventures, both of us trying to escape our teen years. I graduated a year ahead of her, and soon after, she dropped out of school. When I found my first apartment, she helped me move into it, but after several days together it became apparent we needed some space. Emotions were running high. Our friendship would inevitably change after that. She was a wonderful person, but needed more support than I was able to give. You know the kind. After a marathon of support from me, she finally left the apartment and said that she would call the very next day. Her parting words, which meant nothing at the time, still echo in my ears today. "Aren't you going to stop me?"

My answer was, "No, you need to do what you have to do." She was distraught as she left, but that eluded me until later. She didn't call the next day as planned, and I didn't call her either. The quiet time and space felt good. Still, after not hearing from her the second day, I reached out. She didn't answer her phone, and she was nowhere to be found. She had vanished. I called the police to report her missing and got the "seventy-two hour song and dance." I was to call back the next day. My own preoccupations got in the way, and the seriousness of my missing friend seemed to diminish. I didn't call back. My best friend was still missing. Was she dead or alive? Why did I let my need for some space interfere with my love for her? Why didn't I stop her like other times? She had always been there for me, even if I wasn't there for her. Another person in my life was gone.

The desire for someone to stay in my life and love me unconditionally continued on. I thought *love* was the attached word. As I reflect back, it was *security* and a feeling of *stability* that I was searching for. To comfort myself, I grabbed at a variety of empty relationships, which left me reeling when they failed. I had just decided to get some order back in my life when RJ came on the scene. We met at the grocery store and he asked for my number. I told him no. Something inside me recoiled. A couple of hours later, a friend left a message telling me that she had given some guy my number, and I was suppose to call her back. It was the same guy. My inhibitions faded into humor. I called and things would never be the same again. I ignored my previous gut warning and ran into a sick cycle of charades and deception. So far, all that had been good had been taken away. The faith I had harbored since childhood had been lashed with waves of insecurity and the need for love.

Now a prospect for change had arrived. I was curious. He was mysterious and charismatic, holding himself at a very prestigious level. He carefully worked to gain my trust. Subtly, he played on my emotions of sympathy. His story was very different from the ones I had heard before, but who was I to doubt it? The odd things he did weren't strange, just not what I was accustomed to. RJ put no pressure on me to give myself to him sexually, which was the only thing that I had continued to control and hold dear. The fantasy of the "one" that would be my husband to share this with was the only part of my little-girl dreams that was still intact. The relationship grew, and with it came more odd behavior. I grew suspicious, but my sympathies played against my better judgment. He kept up the appearance of a businessman. There was something about his style of independence that was attractive and seemed stable. I had something in my life again, something different.

One evening, after dinner, the romance jumped up a notch, and I didn't know there would be no stopping what was to come. The respect he had shown before seemed to disappear, and my pleading fell on deaf ears. He took what he wanted, and in disbelief and fear, I let him. I would soon discover how much he had played on my emotions and had calculated his actions to weaken my resolve. No sound was coming from my body, just the overwhelming, gut-wrenching feeling of guilt that I had just lost part of myself. It was done. He seemed proud, and ignored the fact of what I had lost, gloating only in his gain. A greater trust had been broken that night; trust in my own judgment and behavior. More than ever now, I wanted out of this lifestyle. Even more, a loss of innocence and a gift that could never be recovered had slipped through that crack of time. In less than a week, the truth would be known and I would be shattered again.

As the detective explained his mission, the true identity of RJ would be unveiled. His life was an intricate woven web of crime and lies that I had flown into. I was the small fly stuck in the trap, and was lucky to not have been completely eaten by the spider. The car he had driven was stolen and I was suspect; the mob ties were unveiled and I was suspect; the parole jumping was uncovered and I had abetted this criminal. I had no knowledge of any of it, but was suspect. The investigation continued and I was eventually cleared of all willing involvement.

An agreement was made with the detective to help recover the stolen property. I would wait to hear from him and would attempt to acquire the information they needed on what was stolen. RJ's call did reveal where the property was, and also gave me the brief opportunity to ask this vulture, "WHY? How could I have trusted you? How could you have taken the part of me that was only for me to give away?" No answers, only excuses and more manipulation. RJ threatened to take his life if I left him. I had no concern for him, only a calloused reply from a betrayed heart saying, "You better do a good job of it when you do it." The detective gathered my information and was grateful. He reached out to me, which was the beginning of my path to healing, even though I didn't know it then. He spoke to me like a loving, caring father and assured me that there was help for me.

Time passed and I was listless. I existed from day to day. At least this was comfortable and familiar to me. With no true friends, an empty void settled inside me. Depression lurked around every corner, and desperation threw me to my knees. I needed help out of this life cycle. I began to turn back to my faith.

Things got worse before they got better. The realization came that I was pregnant! Time seemed to stand still that day. Everyone moved about, the cars went up the avenue, the clouds speckled the sky, but I was alone. I needed my parents, but how could I tell them? They taught me to know right from wrong, though some issues were never discussed. What would I do? I told the events of the unborn baby to only one friend. After talking with her, the seed would be planted to terminate the pregnancy.

The time had come to return home. Another friend, also in a crisis, would go with me. I was still in pain, and was anguishing over the fate of the unborn child, as well as the devastation this news would bring to my family. After a misunderstanding with my friend, we were in the heat of an argument. Our tempers flared as we moved down the hall, and words were exchanged. In a very cold and callused way, she betrayed me and divulged my secret shame to my parents. My father, sitting in the living room and watching TV with headphones, saw the exchange. Concerned, he removed his headphones and asked us to settle down and explain what was going on. She blurted it out. "Go ahead and tell your parents. Tell them you're pregnant!" My father's face froze. He quietly got up from his chair and went to find my mother. It was late and I heard him tell her of my pregnancy. She sat up in bed and cried out to God. The air was thick, and bottling up my anger, I gave way to my shame. The moment of silence soon turned to tears. How could I have done this to my parents? My mother and father sat with me for the next forty-five minutes as I shared with them the details of the relationship and the pending birth. The baby would be born out of wedlock, and it would also be biracial. As I listened to myself telling the story to my parents, I was actually processing the entire chain of events for the first time.

My father stated that he was struggling with the thought of a grandchild. Resolute in my decision, I told them, "There will be no grandchild. I've already decided to terminate the pregnancy. I will not bring a baby of 'date rape' into this world." My words were like a blanket being thrown over a smoldering flame. There was no visual movement and the sound was gone. The guilt and shame was too much. The next few days were complete torture. The family went about in a state of shock. No words were exchanged other than normal pleasantries.

The third day after the whirlwind brought the first words of recognition from my family. My devoted, God-loving father placed his hand on my shoulder and let me know, in a sympathetic way, that he and my mother would support me no matter what I decided. There were no other options provided; the choice was mine. This was the first time I felt unconditional love from my father. However, this unwanted pregnancy would ultimately pull my family apart.

I held fast to my previous decision and the procedure was carried out. The baby was gone. My parents had dropped me off at the clinic, and a short time later, retrieved their little girl. We would live for years with the "code of silence." The events of that time and season would not be discussed. It was their way of moving on. My guilt, fear and torment would be swept under the rug, and would not be shared with anyone, at least not for a while.

I would be on the road to new beginnings soon. Winter came and went. Spring held a promise of new life. A creative season offered some reprieve from old memories, and I began to live again. Still, the reality of taking a life could not be dismissed, and would often come up to the surface of my heart, only to be pushed away by my next activity or duty. I became a master of disguise and learned well how to survive. I tended to others in need, lending them something of value for their lives. Oil painting also became a passion, and when lost in that distraction, I began to have small glimpses of healing. It was my past and its demons that I could not face. Revealing my feelings was considered a weakness, so I hid them well with just a simple flick of my "don't let it show" attitude. I was pretty good at it, but I longed to be free.

My new friend, and future husband, knew nothing of the anguish that was still inside of me. Though I shared my past with him, neither of us saw the depression that was lurking ahead for me. There was work to be done in my life, but I had not stopped and asked anyone for help. I didn't trust many people. I soon found that I had to return to God in order to be healed. I had to renew the belief and trust that was once familiar, and reap the love and joy that was waiting for me all along. Wedding bells tolled about a year and a half later, and we enjoyed married life. Within a year, we moved to Cle Elum, where I focused on our new home, making things work between us. We found a good church home, and God started working within me. As new friendships began, I was surprised to discover how quickly I shared my past. With every disclosure, something similar to a healing balm poured over me.

My husband and I continued to grow in our relationship, but expectations were far from what I had hoped for in my dreams. My anger began to find avenues for expression, which hurt our marriage. We would fight terribly at times, resurrecting old insecurities that dictated much of what I did. The trust I had placed in him was broken, and I started shutting down emotionally. It was very subtle. Early signs of anxiety and emotional stress began to be felt by me. Our empty nest seemed to loosen the marriage ties, and a deeper fear of family and pregnancy set in. God would not allow any of that to remain, though. He had a plan of restoration.

It was in the delivery room with my best friend and her husband where I was finally able to grieve the loss of my aborted baby. They were so excited about the birth of their child, and I had killed mine. That truth pierced my heart. I rolled around the corner, slid down the wall and tears began to flow. Those tears continued on and off for the next month. In memory of my baby, I planted a rose that another friend had given to me. Ironically, it was a Peace Rose, and it bloomed for two years. To my surprise, the next year the wild rootstock took over and it turned into a blood red, climbing rose. Now it was not only a symbol of my lost child, but it was also a symbol of the blood of Christ that covered all of my sins. I dug up the rose and moved it. The plant survived my abuse and the taproot sprung up to new life. Now there were two beautiful plants. This plant would not die. This was God showing me that life does grow and multiply, and that new life can come from old roots after all.

Shortly after this first bit of healing, my husband and I had an opportunity to go with a mission team to a Ukraine orphanage summer camp. God opened yet another small door for both of us. It was something that we could share together and we loved it. That trip spawned a deep love for these children. God would use this trip to usher in more healing and restore a maternal desire that had been lost years before.

With more healing in place, I began to welcome the prospect of having our own children. I waited and it happened. I was late with my cycle! Could I be pregnant? Christmas was right around the corner and a vision of booties and a bassinette was dancing in my head. Slight physical changes were occurring and I couldn't keep it a secret; it seemed to be true. I told my mother-in-law of the possibility, and her joy was inexpressible. It was settled, or so I thought. Another week would reveal that it wasn't so, and the disappointment would sink into its all too comfortable spot, right in my heart. Another loss right after the door had opened. Was the sequence starting again? I started to focus on anything and everything. I had a million plans, and never began or finished any of them. They were convenient distractions.

Our love for the Ukraine continued and another visit was set in motion. This time, they were coming here! Two Ukrainian girls won our hearts during their stay with us, and we entertained the idea of adoption. I was emotionally engaged at this point, and passionate about this pursuit. Those dreams were quickly doused by the reality of the country's politics and enormous adoption expenses. Another disappointment, another setback, and this time it had its hooks in deep.

It wasn't obvious at the time, but I was running out of distractions. I could feel the gravity of its pull wearing a new disguise. Temptation came in the form of a fellow coworker, and it was emotionally paralyzing. I didn't have a file for this one. It would pound like a hammer, driving me into a deep, deep pit over the next two years. A physical relationship never emerged, thankfully, but the familiar torment of guilt and shame were after me again. Despair began to position itself, and the depression went deeper than I have ever known. This time, I couldn't keep busy enough to outrun it. Home became my prison. Many days would be spent lying on the floor in the fetal position, crying until my skin would burn from the tears. My husband would come home, grab a chair, spin it around, and ask me how my day went. Each time, he listened to my story as if he were hearing it for the first time. God sent this man to me, and he never left my side. My husband watched as the physical side of the depression set in. I experienced pain in my neck, dizziness, and a panic that left me speechless for moments at a time. This scared both of us. He was feeling my pain, and his love and support never wavered. We both knew we were in a serious battle. I doubted everything and almost everyone. The validity of my faith was being deeply tested.

God's healing was being put in place, but I remained unstable and afraid. He didn't give up and continued to lay the groundwork. As I crawled up from rock bottom, He, along with my husband, would ultimately support and hold me. It took time to learn to stand gloriously, as I stepped out of the fire, with the grace that was waiting for me. Simple prayers became a ritual, and Bible verses would speak words of promise that became my spiritual food. I had to learn to think differently, love purely, forgive and let go of anger. Most of all, I had to begin to trust the love that was offered to me. I soon found out that my husband truly was a gift from God. God placed him in my life to love me unconditionally, and show me that I would not lose him, as I had lost others. God continued to work in me. In response to my desperation for answers, God assured me that He had allowed this season to come because it was the only way I would listen. The rejection I felt from others was not because those people left me. Rather, God had removed me from them, so that I could spend time alone with Him. Also, I needed to let go of the taskmaster view I had of God, accepting Him as the true lover of my soul. This dark season was appointed to teach me what love truly is, and to deliver me from the past. It was the time I needed to learn valuable lessons that would prepare me for the near future. From that moment on, it was baby steps. We were on the road to recovery. My husband and I renewed our commitment to each other, and made a vow to not view each other as the enemy. We had a desire to see what God could really do in healing our marriage, if we would let Him.

In 2004, God took me back to the Ukraine. What a homecoming that was! The trip rekindled a love that hadn't slipped, even after all those years. It was a time of more healing and further testing of my love and devotion to both God and my husband. That love was proven genuine. I returned from the Ukraine with a promise inside: there is a living hope in this life that cannot be spoiled by what happens around us or to us. For the first time in a long time, I truly knew joy and peace. I received a promise that things were going to be radically different in two years, so start getting ready!

One more mission trip to the Ukraine in less than a year left me starry-eyed. What followed is God's gift to us today. It brought me to a new place, a place of acceptance, and the beginning of a new cycle of hope instead of defeat. God was preparing me for more changes with words from the Bible, to challenge and heal all remaining doubt.

In Hebrews, Chapter Eleven, the Bible tells how, through faith, Sarah received the ability to conceive. The possibility of conception was all but dead to me. Yet, God had done so much healing, how could I doubt this? I believed that He *could,* but was hesitant to believe that He *would.* As I read deeper in Scripture, it tells that women, by faith, would see their dead raised to life again. The "dead" were children they had lost. This meant something to me. My walls of disbelief were shaken and I was being tested. Now I had to choose. Would I stay and be crushed in disbelief, or accept the truth that all things are possible with God? I learned that it takes faith to welcome unplanned things, and faith to receive back life. Having no idea where it would take me, I desperately wanted to believe and have this faith.

That summer, we made our regular trek to Montana. My husband and I really had made it through the fire. We had been burned, but not consumed. It was a new season. We returned home and I began to once again notice some physical changes in me. My previous breakdown had caused my cycles to be completely unpredictable. We talked about it and waited. Fear began to tiptoe around, and that same old doubt began to pitch fastballs at me. What was going on? It seemed easier to accept some other physical problem than the simple prospect of being pregnant. It had to be a problem with my system. So, in the process of elimination, I took a pregnancy test, and it was negative. The debris of disbelief was falling again. I scrambled emotionally to stand strong, wanting faith to win the day. We waited and delayed another pregnancy test to avoid going through the pain of another false hope. My life was heading in a good direction. How could I take a step backward now?

On another Sunday, a sermon was given that told of a woman who had been very helpful to a certain prophet. He repaid her kindness by telling her she would be holding a baby boy in her arms this time next year. I was overwhelmed by these words, wondering, "Was God speaking to me?" It took time for my stubborn countenance to accept the simple truth. In a conversation with my friend, the pressures seemed to build for the truth to be acknowledged, so I screamed, "Okay! I'm pregnant!" We both laughed with relief. My husband was thrilled and excited when I was able to tell him with absolute certainty that we were going to be parents.

Now, here we are, eighteen years after that U-Haul truck pulled away. I'm alive and I've made it through the fire. God has delivered me from the worst. My husband loves me unconditionally, and he is my best friend. He has a devoted, loving and grateful wife that has the utmost trust and respect for him. We also have wonderful news: *God has created a new child that will come to Him through me.*

There is more to our journey. With God as our guide, my family will continue to walk in His name. Praise God for His healing and His forgiveness!

Editor's Note: As this story was going to print, a healthy baby boy, Nathaniel Gregory Greenman, was born to Rob and Miriam Greenman on March 14, 2006.

7

Turning Tragedy into Triumph

The Story of Gil Ward
Written by Susie Weis

As the officer was shining his flashlight around to survey the damage from the accident, I looked down and caught sight of a shirtsleeve. Then we saw it, an arm protruding out from under the bent wreckage of the car. It was Sam.

I had just arrived home after being discharged by the Army, and was out to party. Earlier, my good buddy, Sam, and I had gone to pick up a '59 Chevrolet Impala that he had bought that day. In celebration of his latest "ride," we asked his girlfriend, Sharon, and another girl, Jackie, to go with us. We were all having a great time speeding down the old highway, consuming drinks of Southern Comfort. The car didn't have a speedometer, so we had no idea how fast we were going. We came upon a sharp corner way too fast, and the next thing I knew, we hit the snow bank, became airborne, and flew over one hundred feet through the air. My life flashed before me in slow motion, like watching a bad movie on a motion picture screen. The girls were screaming hysterically, glass was breaking, and everything was being thrown about. Somehow, I flew clear of the car, but the girls were still in the car and hurt badly. I couldn't find Sam, and it was too dark to see without a flashlight.

"Where's Sam?" I thought to myself. "He must be hiding to avoid getting into trouble." Someone who witnessed the accident hurried down to a local gas station to call the authorities. After what seemed like an eternity, the police finally arrived. While we were checking out the scene, and I was trying to explain to the officer what had happened, we found Sam. He was my best friend, and he was now dead. I will never forget the sound of crunching metal and the loud screams. That day will forever be seared in my mind. God had taken my friend away, and I was completely lost. Up to that point, my life had been pretty much a living hell. Sam's death just added to the pain.

Pain! Anger! Resentment! These are just a few of the negative words that defined my childhood, adolescence and early adult life. My pain began as a little boy when my step dad, Dick, would beat me just to take his anger out on someone. At four or five years old, I was an easy target. There was no defense against his rage. Anything he could get his hands on, a curtain rod, willow switches, the back of his hand or a fishing pole would work just fine as a tool for my beatings. In fear of being abused herself, my mother would do nothing to prevent my step dad's cruelty.

I just wanted to be loved and to be close to my step dad, but nothing I did seemed to please him. There were times when he would get up and say, "Hey, boy! Do you want to go to the rail yard with me today?" Trying to always please him, and so excited that he wanted me to be with him, I would rush to get ready. It was going to be a great day and I didn't want to do anything to upset him! I would think that things were finally going to get better, but I was always wrong. At any time, the switch would be flipped, and when I least expected it, he would get mad over some small thing and I would be the target of his beatings again. I was convinced that I was my step dad's worst enemy.

In addition to the physical and mental abuse from my step dad, my older cousin, James, added to my pain by sexually abusing my sister and me. He would make us go on walks with him, taking us to secluded places and then forcing us into sexual acts with each other. While he watched us in our moments of shame, he would find his own disgusting pleasure. He would threaten us, saying, "If you dare tell anyone, I will say that the two of you always started this. Who do you think they're gonna believe, someone older like me, or two little worthless kids like you?" He'd then tell us, "Your step dad will beat you to a pulp if he ever gets wind of what goes on out here." In spite of his threats of a beating, I tried to tell my mom what James was making us do. She didn't believe he would do this to us, and chose to not make a scene. Thus, the pain continued.

I felt so worthless. It was easier to just withdraw. All I knew was that everyone who should love me and care for me would instead hurt me. Anger was my constant companion, and building walls around me seemed to be the best way I could cope with my life. If anyone tried to get close to me, I would wonder, "What are they going to do to me? How are they going to hurt me?" I would think to myself, "Am I ever going to find anyone who really loves me?" Instead of love, rage filled my heart. My mind became consumed with thoughts of wounding or killing those who had hurt me.

My mom finally left my step dad, and we began moving around a lot. She ended up with many different men. Most of them were very abusive to her, which seemed to be the norm in our lives. It was during one of these moves that I met a man named Les Roberts. I felt he was an angel that had been sent just for me. This kind man took me under his wings. We spent lots of time doing those things boys should do with adults: playing baseball, going to movies and eating at fast food joints. He was a paratrooper and would often take me to watch him jump from the plane. I will always treasure the pilot chute he gave to me as a memento of our times together. I had finally met a man who treated me like a friend, who wanted to be just with me. It gave me hope.

Our next move was to Cle Elum, Washington. Wanting to be accepted, and looking for a father figure in my life, I started hanging out with guys who were a couple of years older. They were always getting into trouble. On the first day of school, I skipped classes with a couple of these older guys and broke into a home. This was just the start of my serious trouble. Along with truancy and robbery, I was drinking beer and hard liquor. Booze was my best friend because it helped to get rid of the nightmares in my head. I had no one to talk to, and thought no one would believe me, even if I did try to talk to someone about the anger I was feeling. I started using marijuana, and eventually turned to hallucinogenic drugs like Orange Sunshine acid, still trying to mask the pain.

It was during this time that a glimmer of hope came into my life. My friend, Vern, and I were entertaining ourselves by exploring in the attic of his mom's house. I spotted an old guitar in the corner that was in pretty bad shape. To me it was the most wonderful thing I had ever seen. It was so intriguing that I asked his mom how much she wanted for it.

In her gruff voice she replied, "I'll take five bucks for it."

"Sold," I shouted. I did chores everywhere I could, until I had earned enough money to pay for the guitar.

I totally escaped from life each time I strummed that guitar. That old guitar's strings were so far off the frets that I had to push extra hard to make them sound like anything, and my fingers became very sore and calloused. My mom wasn't too thrilled with my newfound passion, but nothing could stop me from playing. I would play for anyone who would listen, and after a very short time, I actually became fairly accomplished. I know now that music was my escape from the pain that kept torturing me. When I was playing, the music calmed me and kept the anger inside of me from always being at the forefront of my life. When I wasn't playing, it continued to fester and get worse.

I still hung out with the wrong crowd. The trouble I was creating for myself got so bad that the authorities gave me two choices: go live with my real dad,

Henry, in California, or join the military. I had never met my dad before, but I chose to go live with him rather than going into the service. Henry and his wife had two young daughters, and it was no surprise that his renegade son was not welcomed there. Because she wanted to get rid of me, my stepmother planted dope in my room to show my father I did not belong in their house, and that I would be a terrible influence on their daughters. He believed her, and feeling totally rejected again, I ran away.

Aimlessly drifting to nowhere, I was picked up by a man in the Air Force. He took me to the base and told me that I could spend the night at the barracks in his room. Feeling physically and emotionally drained, I welcomed the shelter. There were two beds in his room, so I crawled into the empty one. In the middle of the night, I was awakened to find him in bed with me, making advances. I was furious, and pushing him out of bed, told him to leave me alone. He knew I was serious. I was only seventeen at the time, but I knew I needed to get out of there. The rage dwelling in me erupted, and while he was sleeping, I stole his wallet, his money, his car and his military ID (thinking I needed it to get off the base). Later that night, I stopped to get some gas, and nervously locked the keys in the car. I was sure he would have awakened by now and called the cops. Instead of taking the time to get the keys out, I left the car right where it was, and found my way back to my father's house. The next day, I explained to the Military Police what had happened, and fortunately, they didn't press any charges

Since it hadn't worked out to live at my dad's

place, the authorities had no choice but to insist that I join the Army. Everyone hoped the discipline and structure of the military would help straighten me out. It did nothing. The military just gave me another avenue where I could act out my aggression when I was barked at, commanded to do things, and expected to respond instantly to an order. Often I would flatly refuse to obey an order. I received an early dismissal from the Army with an Honorable Discharge, however, my DD214 release papers state: "Discharged for the good of the military."

It was on that fateful night, just after being released from the Army, that my friend, Sam, was killed in the car accident. After that, I really fell off the deep end and was hell bent for revenge. There was no stopping the pain. Booze and drugs, hash, speed and cocaine, had become my only friends. Without a purpose, lost and alone without any structure, I drifted from place to place.

In Cle Elum, I met my first wife, Fern. She was a very sweet lady who deserved much better than me. We eventually went back to her hometown of Clarkston, Washington, where she found out she was pregnant with our son. Neither one of us was ready to be married, let alone become parents. She did her best to create a good home, while I, on the other hand, continued to do what I wanted to do when I wanted to do it. I was like a Dr. Jekyll and Mr. Hyde. I would be really sweet one minute, and then blow into a rage if she questioned me about anything. I was never interested in what she thought or had to say. It was either my way or no way. I thought nothing about leaving home for weeks at a time. When my lead guitarist and I decided to take a little vacation to Alaska to play our music in bars, I didn't even think about involving her. I just did my own thing and left. My whole world revolved around music and taverns, not her. I lived out the typical musician's life: drinking, partying, meeting women and having sex. One day, while the band was setting up to play, a state trooper walked up and presented me with divorce papers. I ripped them up, and thought, "How dare Fern do that to me!" As far as I was concerned, everything was fine in our marriage and I certainly refused to take responsibility for anything that had happened. My own self-destructive behavior had destroyed our marriage.

Singing and playing the guitar were things I loved and did well. I had no specific plans for my life, but I knew music was something I would always do. It was a way for me to relieve some of my stress. The band played every night in the bars, and we would drink until dawn. Life was like a bad merry-go-round that was spinning out of control, and I couldn't get off. At one point, the vicious cycle got so bad that I got drunk and started taking pills to kill myself. I guess I really didn't want to die, but was instead just looking for relief from my pain. Fortunately, I had enough sense to call someone for help. I was taken to the hospital where the doctors pumped my stomach and kept me overnight for observation. I was told that I was an alcoholic and needed to get help. I wasn't ready to give up my drinking, but agreed to attend the Alcoholics Anonymous meetings. I didn't really want to cure myself, but I knew I could play their games and go through the motions which would satisfy them. It wasn't long before I was back into boozing heavily. I was still lost and in pain, with no one to rescue me.

It was about this time I met my new wife, Annie, who caught my eye the very first time I saw her. She worked as a bartender in Homer, Alaska, at the Bayside Lounge. I was running around with one of her friends, so Annie assumed there was something between us. When our gig was over at the Bayside, and we were packing to leave, I noticed Annie standing by the door to the lounge. I said, "Hey, maybe next time I'm in town, we can go out." She politely informed me that she was happily married and not interested in dating. I responded, "Well, if you ever become unhappily married, let me know."

Some time later, while performing again at the Bayside Lounge, Annie told me her happy marriage had become an unhappy one, and it had ended. That evening she had invited some friends to her house, but they were unable to make it. She then asked me if I wanted to come over for dinner and watch a movie. The evening was fantastic, and from then on, we spent as much time together as possible. It wasn't long before I realized she was that "special someone" with whom I wanted to spend the rest of my life. After our gig at the Bayside had ended, much to my disappointment, I had to move on to another town. One day, a mutual friend of ours called to say that Annie had really been missing me, and even thought she might be in love with me. On the spur of the moment, I told the girl to get Annie to the airport, a ticket would be waiting for her at the counter, and that I would meet her here when she arrived. Five months later we were married!

But the happiness of finding Annie didn't stop my alcohol consumption. While playing with the band, I always drank, and as soon as I awakened in the morning, I started drinking again. It absorbed all my days and nights. Just before I needed to get ready to play again, I'd take a short nap, put Visine in my eyes so I wouldn't look drunk, and then head off to the bar. This went on for many months, and I got very good at not showing how intoxicated I really was. Even though drinking helped to numb my body, I still felt the pain of the past that continued to haunt me.

One day, I was walking the streets of Cordova, Alaska, where we now lived. There was such emptiness in my heart that I just couldn't quit crying. My life was completely out of control and I didn't know how to stop drinking. As I was walking and crying, I was drawn to a church. There was beautiful, soothing music coming from inside, and it touched me in the innermost parts of my soul. It was a pleasant day; the sun was shining, and eagles were flying high in the sky. I could not understand the complete emptiness I was feeling. The closer I got to the little church, the stronger this feeling of peace came over me. I started weeping uncontrollably. Since the door was already open, I walked into the church, and right in the middle of the service, I shouted, "I need Jesus!"

The congregation of this little church immediately stopped the service and prayed for me. I could feel real love coming from them. They showed me that God loved me and that He really cared about me, as well as all the pain I had experienced in my life. I wasn't just a crazy drunk that stumbled through the door, but someone who was in need, and they were there to help. These people knew Jesus personally, and they took the time to share Him with me. When they asked if I wanted to ask God to come into my life and to forgive me of my sins, I immediately told them yes. I prayed the sinner's prayer and accepted Jesus as my Lord and Savior. I knew this was what I had been searching for since I was a little boy being abused by my step dad. Because of Jesus' love, all the pain I had endured in my life began to heal, and my anger and frustration turned to joy and peace.

I was so excited to tell Annie about what had happened to me, and how Jesus had saved me! To my great disappointment, she was very angry and didn't want to hear anything about this new life I was living. She wanted her "partying" husband back. She knew me as the musician who lived on the wild side and was constantly drunk, not someone who was continually talking about Jesus and trying to convince her she needed to be saved, too. She was furious, and yelled, "Don't be telling me any of that Jesus stuff. Just leave me alone. Go preach to someone else!"

I couldn't leave her alone, and I never gave up on her, I loved her too much. I prayed for six months that she, too, would come to know the Lord as her friend and Savior. One night it finally happened. After I had worked a very late shift, I heard Annie coming up the stairs to our place, returning from a bar. She was singing loudly, and was, obviously, very drunk. The Lord prompted me to get up and watch her. As soon as she got into the house, she went to the kitchen and grabbed a knife, holding it against her stomach. She was going to kill herself. I was able to maneuver myself around and take the knife away, but not until she had cut her finger slightly. She sat down on our bed and watched the blood drip onto the sheet. I heard her speak in a strange voice saying, "I want to die." It really scared me and I knew I needed some help from our pastor, who lived nearby. But just as I was opening the door to leave, I realized Annie might stab herself with the knife if I didn't stay with her.

I cried out, "In the name of Jesus, I command you to leave her alone." She fell back on her bed, and then sat right up again, as if there were a spring that pushed her upward. She was laughing in that same, strange voice. I took my Bible and again commanded whatever was controlling her to leave her.

She then said softly, in a small child's voice, "Can you read me the 23rd Psalm? Jesus is here. Jesus is here for me."

With my Bible in hand, I quickly began reading that Psalm to her. It wasn't long before I realized we were in the presence of the Almighty God, and He was there for us.

I said to Annie, "Do you want to ask Jesus to come into your heart?"

She tearfully answered, "Yes! Yes, I do."

As she gave her heart to the Lord, I had the wonderful joy of leading her in the sinner's prayer. All the sins of her past were removed. She now understood the joy of knowing Jesus and that He had forgiven her sins.

After that, our life together began to change in miraculous ways. We started attending the little church on a regular basis. Learning about forgiveness allowed me to start to understand the cleansing that was happening within me. If I could be forgiven, then I needed to forgive others for the pain they had caused me. I was able to put the past behind me, and started looking towards the future, knowing that God would always be with me. Life was still a struggle, but we learned to serve the Lord and look to Him for direction. I began to find hope instead of despair and pain.

When my step dad was visiting my sister, I called and talked to him for the first time since mom and I had left years ago. I knew I needed to share with him what Jesus had done in my heart and my life. Action speaks louder than words, and it was time for me to put action to my faith. One way for me to do that was to tell Dick I forgave him for what he did to me as a boy. At first, I was anxious about having any conversation with him, and the possibility of resurrecting the old pain he had caused. However, when I heard his voice, the Lord removed all the anger and resentment I had harbored for many years. I was able to say, "As a man, I can't forgive you for what you did. But because I know of God's love and that He has forgiven me, I forgive you." Shortly after our phone conversation, Dick died from cancer. I was so grateful for the opportunity I had to speak with him about God's forgiveness.

My mom, who was living in Cle Elum, became very ill. For many years, she had been smoking several packs of cigarettes a day and was now diagnosed with lung cancer. One of the lobes of her lung had to be removed, so we hurried down from Alaska to be with her during and after surgery. While staying with mom, we checked out several of the churches in the area and found a local church that made us feel at home. It reminded us of our little church in Alaska. Feeling the direction of the Lord, Annie and I returned home, packed our things, and moved back to Cle Elum. Because we knew we had found a new church family we could love, it was so much easier to say goodbye to our wonderful friends in Cordova. Things had really changed for us in Alaska since getting saved, but we still had our old friends and my life of music. This move, which took place about ten years ago, gave us a new lifestyle and helped us with a fresh beginning.

We became members of that friendly church and are still very active there today in the ministry of the Lord. I am able to use the gifts the Lord gave me by playing the guitar, singing for His glory, and helping to lead the worship services. I have more pleasure and joy singing God's praises than any other music I have sung in the past. Annie provides assistance to the church and the community in the area of "helps and hospitality." She also supports local young people and juveniles with their spiritual needs. We are both licensed, ordained ministers, and our goal is to be a lighthouse to this community, sharing God's love with others.

There are many turns in the road that I have taken; some I would not want to change, and some I would want to do differently. I know I am a better person for suffering through this journey. I have learned that God gives us talents for His good work. If we think we have a better idea than God, we may have to play the fool until we let Him lead our lives. Most importantly, I know that no matter how many times you try to escape the pain, it will only follow you and torture you. The only way to let go of the past, and all of its problems, is to walk in the Lord's footsteps and be washed in Jesus' blood. Then you are truly free.

The pain...anger...resentment? It's gone because of God's great love. It has been forever replaced by a joy in the Lord I never thought possible.

8

A Passion to Win

The Story of Mike Miller
Written by Miriam Greenman

The track lay like glass before me. It's a familiar ritual of sitting in the seat and waiting for the starter motor to spin the 3,000 horsepower "nitro" motor to life. With the seventeen-inch-wide Goodyear slicks under me, it was time to make the burnout and heat up the track for a run. I've been passionate about racing since I was a child. As a teen, the Winter Nationals and Don Prudhomme, drag racing's star, were my study hall fantasies. As I looked through magazines and read articles, I wanted to be there, like them, going fast!

My passion to race turned into a passion to win, and now my dreams from study hall had merged into reality. My heart was pounding; we had made it to the finals! I had eliminated the likes of "Jungle Jim" Liberman, Tom "The Mongoose" McEwen and Danny Ongiasis. Only Don "The Snake" Prudhomme was left to beat in the final run of the Winter Nationals. Beyond any win, this was a dream come true. The driver's seat was now a comfortable home to me. The engines thundered and the smoke began to roll in the staging lanes. Memories, in slow motion, crashed like a tidal wave as the "Christmas tree" (starting light) counted down. Suspended in time, those bygone days as a boy came rolling in.

They raced old cars or jalopies back then. That

was the name of the game. They emerged from nearly every garage in town to gather at the Sunset Speedway, nestled in the foothills of the Cascades. Frankenstein in their construction, these crude cars were the race cars of the day. With 110 plus horsepower in rein and bandannas at half-mast to keep out the dust, they ran many races on the old dirt track. I can still hear those old jalopies rumble, and remember vividly the first time I watched them with my Dad. This passion to race was injected into my blood. It was all about racing.

I grew up in Cle Elum, Washington, where my family owned the local sawmill. I guess you could say I was considered a silver-spoon kid. My first exposure to racing was when we would go up to the track in Easton on weekends and watch the old jalopies race. It was there that I caught the bug for speeding and competition. I fooled around with cars through my teen years, racing on the weekends and stirring up trouble. It's what we did back then. After high school, I joined the Air National Guard and put racing on hold. As soon as I finished basic training and tech school, I was ready to pick up where I had left off...racing. I heard about a new craze called "Funny" cars. They always drew a large crowd and came with a big paycheck, if you just showed up with the car. It was irresistible!

My first car was a 1968 Dodge Dart, a fiberglass-bodied, Chrysler-powered "flopper", the term used by racers for Funny cars. I raced locally for a while, but it wasn't long before the tracks called me south and I landed in sunny California. The months in California were a dream come true and passed like fuel going through a carbure-tor. We had a nationally-ranked car and we were win-ning. Eventually, I ended up as a hired driver, and on the starting line in the finals at the Winter Nationals.

Every win has the shadow of agony in its back-ground. Prior to the NHRA (National Hot Road Associa-tion) Winter Nationals, we were at the US Nationals at INDY, the Madison Square Garden of Drag Racing. We had made some changes to the car, but still hadn't quali-fied. It was our last chance before dusk, and we were in the last session of the last round and on the last pass. That's when it happened. Orange flames engulfed me and I was caught in a 200 mph ball of fire! The car had kicked a rod and burst into flames, and I was in it! In an urgent reflex, I released my grip on the wheel, hit the three extinguishers, and pulled the drag chute to get the rolling flame to come to a stop. "Wide World of Sports" caught it on tape and forever immortalized it as "The Ag-ony of Defeat."

We had all of Saturday night to put the racecar back together again. It was another unpredictable chal-lenge, but, like a race, we weren't defeated yet. Everyone worked hard replacing the engine, drag chute, and tires, to have it together for the semi-finals. Sunday came and we went two rounds at the U.S. Nationals before losing to "The Snake."

We continued to run the circuit, but once again, I got the itch to have my own racecar and crew. In those days, we would go back to the East Coast to compete in match races. We would be gone from April until early

September. The tracks were close enough that we could race up to three times a week and actually make a living.

By the time the eighties came around, things began to change. Gradually, racing became more corporate. The economy was shifting, and the tracks were beginning to thin out. Money wasn't there to support private cars and crews. NHRA became the only viable outlet for racing. To stay in the game, you had to scramble to find a money sponsor. It was some time before we finally landed one. I still had my crew to support, and we needed to get ready for another season. This would take more than a small amount of money that you could float on a personal loan. My wife, Shirley, and I were already struggling with our marriage; the strain of having no money and my never being home was taking its toll. Shirley knew something had to happen, but didn't know what could be done to change things.

That's when I met Jerry. I had raced on my share of tracks, you might say, but no racecar had prepared me for the next quarter mile. We had been working on the car, trying to hammer out some piston scuffing, when Jerry's son told me that his dad would really like to talk to me. The car was commanding most of my attention and I really didn't have time for chitchat, but I agreed and we met. Clad with a "Jesus Saves" belt buckle and carrying a bottle of Slick 50 oil, he was truly one of a kind. I had no idea what was about to happen. He was asking all sorts of questions about the car, more than the average person would ask, and he certainly knew what he was talking about. He was promoting Slick 50 at the time and

I normally would have blown the guy off, but I was busy and desperate. Jerry had a way with people, and talked me into letting him come to a test session that was scheduled for the next week at Seattle International Raceway (SIR).

I agreed to pick him up in a church parking lot. Surprisingly, I never put the "Jesus Saves" belt buckle and the church parking lot together, to draw any conclusions. I had other things on my mind, like getting to the track and getting the car ready for the test session. When we arrived at SIR, my crew chief and crew were already there. Before the test runs, we huddled around the car, making certain it had no mechanical problems, and certainly not thinking much about our "guest" or what we said.

When the test runs were over, my wife and I agreed to drop Jerry off in the church parking lot. On the trip home, my wife's curiosity got the best of her. I hadn't been able to answer her questions about Jerry, so she decided to ask him directly who he was and what he did. Jerry answered, "Well, I was an engineer, and I graduated from the University of Wisconsin. After graduation, I came out here and went to work as a mechanical engineer for PACCAR. That's my background, and it's always what I've liked to do." Now the pieces started falling together for me, and explained his questions and genuine interest in racing.

Still curious, she asked, "But, what do you do now? Is the Slick 50 oil that you're selling a PACCAR product?"

Hesitantly, he answered, "If I tell you what I do, it may ruin our relationship."

Puzzled, she pushed him even further. "Now, you know, Jerry, there's no stopping here!"

"No, I guess not," he smiled. "I'm the pastor of the church where you picked me up."

I sank low into my seat. Every word we had spoken at the tracks came back to me and I was feeling extremely uncomfortable. As we pulled into the church parking lot, my heart was pounding and the clock was ticking in my mind. Like so many races before, we were at the proverbial starting line, waiting and watching the Christmas tree countdown. However, this was a different race. The driver's seat had always been a comfortable place and now it felt like the hot seat. Like those orange flames that had engulfed me at the US Nationals, I was caught in the car with a 200 mph ball of fire, Jerry. I tightened my grip on the wheel. In an urgent reflex, I wanted to hit some kind of extinguisher, pull the chute and get this rolling flame to come to a stop. But I was stuck. People started showing up. Guys my age and younger peered into the window to say hello to their pastor. That added to the pressure. I just wanted to hurry up and get this over with, lay down some track, heat up the tires, get off the starting line, and head for home. But something happened, and I would never be the same. It was like throwing a rod, and my life was about to burst into a ball of flames. Jerry asked, "Mike, do you believe in God?"

It's a natural thing to get nervous when someone

asks you that question. It's really kind of odd. It's not that I didn't believe. I had always believed in God. Like so many others I knew, I would offer up the famous "foxhole prayer" whenever things were going bad, and then make promises to be good from here on out. I wasn't a churchgoer, and I didn't hold on to any particular religion. In fact, I thought most of them were cults, and I sure didn't want to get into something weird. Much of what I had witnessed up to that point just looked plain kooky. So I guess you could say it's not that I didn't believe, but that I didn't know how to believe or whom to trust. At this stage in my life, however, I was more receptive. At least I didn't run religious people off. I was willing to listen. I thought that religion could be another tool I could use as an asset, and I would be the benefactor. After all, everyone could use a little extra help in their corner to get them through their tough times. For me, it was about winning.

Jerry presented the Gospel to me, and told me how Jesus had died on the cross for my sins, and *if I confessed with my mouth and believed in my heart,* I would be saved. He then asked me, "Mike, are you ready to do that? Do you believe in your heart that Jesus died on the cross for your sins? Are you ready to be saved?" Frankly, I was ready to get out of there, and would do or say anything just so I could! He was very good at overcoming objections and I finally agreed to pray with him. The three of us prayed together, and I made a confession of faith. I would eventually discover that there is more to faith than confessing with one's mouth. I also had to believe with my heart, but that would take some time. I didn't understand what it really meant to believe with one's heart, so I guess you could say that Jerry and I just agreed to agree.

The next season came and I had a whole new plan. I had the idea that God had the perfect place for me. I was to be His ambassador on the racetrack and to win everyone to the Lord...and, of course, win all my races. Yeah, right.

It was the worst season we ever had. We didn't win a single race, and to top that off, we started losing corporate sponsors. At the time, Olympia Beer was one of our major sponsors. No one could have forecasted what would transpire in those next few months. Olympia sold out to PABST Beer, the plant closed, and the workers were locked out. Kenwood was another big sponsor, but the company closed. For about three months, I chased around, flying here and there, trying to put something together. We were losing money fast. After some of the dust of change had settled, we approached PABST to come on as our sponsor. We talked and they listened. They had factories on both coasts and we were going to race on each. The plan was to have two car bodies, one for each coast, and our sponsors would be PABST for the East and Olympia for the West. They had committed one hundred thousand dollars, which was going to get us out of hock. So, we kept traveling, staying in the cheapest places, expecting things to be cleared through Corporate. Finally, the awaited call came in from PABST, but the deal was dead. I couldn't believe it! There I sat in that hotel room, no longer on the edge of my seat, as in so many races. No, I was slipping out of the seat that had been so comfortable, and losing my grip on the wheel that had been so well-fitted. We were truly in the last session of the last round and on the last pass, and those orange flames were coming back, as my dreams were going up in smoke. We were dead broke and it was over. Life, as I had always known it, was done.

The reality of Jesus being in my life hadn't made one bit of difference up to that point. At least, not in the way I had expected. It was still all about the car, about racing, and about winning. I was disillusioned, to say the least. I knew that the whole racing scene wasn't what it should be, and I wasn't everything God wanted me to be. I instinctively knew that, but I'm not sure how I knew. I didn't understand where Jesus fit into my life, but I did have more of an awareness of Him. I wasn't into reading the Bible or praying. In fact, the only time I spent in prayer was when I felt bad about something I did or I wanted to better myself. It was always a negotiation. "If you can make me a superstar, God, I'll do better. I'll shine for you." However, when the situation passed, so did my promises. It was all self-focused. It was all about me!

Truthfully, that first prayer with Jerry might not have been very sincere, but it wasn't without effect. My life was never the same after that. The more I learned about God, the more I learned how to believe. Over time, I must have said that prayer of confession a million times just to make sure it was real!

In retrospect, what God did for me was to remove me from something that I didn't have the courage to end myself. There was no way I could remain in racing and be the kind of person I needed to be for God and my family. Back then, I would have rejected that thought with a passion, and I wouldn't have come to terms with reality. It's not that racing was evil. There are some in that field that are strong enough to get over their egos and do what they need to do. I just wasn't one of them. I was addicted

Letting go was one of the most difficult things I've ever done. I was a very miserable person for about two years. I hadn't realized how the racing lifestyle consumed everything about me and how I was trapped in its clutches. It took a while to relearn how to live. The whole process of letting go was a rehabilitation of sorts. Racing was a drug, and I had been addicted. Slowly, I was able to give up the desire to race, and began the process of reshaping my life.

I started by getting a new job, allowing me more time with my wife and family. The previous lifestyle had taken such a toll on them all. I was working with a road construction crew, laying down fabric membrane on the road before they paved it. I also fixed the cracks. It was ironic. So many sections of the highway were very familiar, but only because I had previously seen them from the seat of a semi loaded with gear. We had trekked this path many times in that semi, and here I was "walking on" and repairing a road that was broken. It wasn't until God took me out of that driver's seat, and put my feet on the ground, that I could see the cracks. It was just like my life. Everything had been set in concrete, and I had been removed from what was actually happening around me. Our lives needed to be repaired. It would be hard work.

We experienced many more changes, as our lives gravitated from one focus to another. Our racing friends that were with us through thick and thin, now started to disappear. We had less and less in common. As close as we had been, our new life with Christ separated us from them. This was a difficult time, but we had come too far and had too much to lose. We knew what was important now.

In life, as in racing, it wasn't long before I got the itch to have my own crew again. I started my own business doing road construction. Our family was healing. There were times on the job that I would reminisce about racing, but I no longer felt cheated because I had chosen this new path. I could really see growth. Realizing that there was a time when I smugly thought that racing was all there was to life, I finally started living for something besides myself. There truly is life after and outside of racing. I often reflect on those seventeen years of racing with regret–so much was sacrificed, with so little to show for it. It was hollow. One day you're a hero, but just like the smoke on the track, the next minute you're gone.

Time passed, and we found ourselves living over the mountain again, back to our roots in Cle Elum. We settled into our new home and found our way into a church family, but it was different from what we were used to. We didn't want to admit preferences, especially at the church level. After all, isn't it all the same? The truth is, it's not. And though you don't want that to get in the way, it can. Soon I was lulled away in disinterest, and I wasn't prepared for how quickly I reverted back to my old roots. I guess, in some fashion, all of us are only one decision away from going back to our old ways. At any given time, it's close, and can happen very fast. It doesn't always have to be an immoral decision. You can just choose to separate yourself from God, and then you're on the road back to your old life. If you're not in His Word, there's no possible way you can stay right with God.

That was another season of growth and change. It's not bad to see your weaknesses; it helps you to see your need for God. I needed Him more, as my weaknesses crept back into my life. It went hand in hand with my racing roots. If I hadn't experienced my success in racing, my dependence on it, and ultimately its failure, I would have lost so much more than I had gained. My weakness drew me back to God. Once again, I found restoration and healing.

I confess that, previously, I didn't know how to believe. I didn't know or understand how it could come from the heart. Today, I know what it means to confess with my mouth and believe in my heart that Jesus is Lord. I'm a different man, not perfect, just different. I have learned to be content.

9

There All the Time

The Story of Rod Cross
Written by Diane Jasper

Just horsing around in the woods after school like all kids do, that's all it was. The neighbor girl grabbed Rod's coat and ran off with it. Since he was big for a twelve-year-old boy, it wasn't hard for him to catch her. He ran up right behind her, grabbed his coat, and started to put it on. At the same moment that Rod was looking down to zip his coat, she raised her hand up over her shoulder, preparing to throw an old knife. In an instant, out of the blue, the blade plunged into his eye. In an instant, out of the blue, death brushed by.

November 18, 1987, was the day Rod lost his eye, changing everything. He was the oldest of four children in a close-knit family, a clean-cut boy who occasionally went to church, and always liked to do what was right. He was good in sports and loved it when his dad came to watch him play. They lived in the small town of Maple Valley, Washington, a nice rural place in which to grow up. He and his brothers and sister spent a month each summer with their grandparents in Chestertown, Indiana, near Lake Michigan. Those months were the good times, enjoying fun-filled vacations at the lake, while playing and horsing around with his many cousins. Rod was a happy-go-lucky, all-American kid.

When that knife struck, stopping only one millimeter from his brain, his happy-go-lucky childhood was shattered. The knife had been rusty and dirty, which caused dangerous infections. By receiving intravenous feedings of antibiotics three times a day, the doctors were successful in preventing the infection from reaching his brain. He spent three months in the hospital, battled painful nerve problems for the next three years, and underwent fourteen surgeries. Life was full of tough and seemingly unfair adjustments.

One of the hardest adjustments was in school. Before the accident and his long stay in the hospital, he was a straight A student and a good athlete. Rod soon fell behind in his classes, and just about the time he was catching up, he would be pulled out of school again for another surgery. His education suffered, and he was unable to catch up in all of his classes. He had always been very good with numbers, but junior high algebra was the last straw, it just didn't make any sense. Rod was very discouraged and decided he needed a fresh start.

Rod convinced his dad he should attend a private Catholic school, where he might have a chance to improve his grades. The first couple of years were pretty good. Rod was a fun guy, and the patch he wore on his eye was a good icebreaker, in the beginning. However, the fresh start he had anticipated didn't materialize, and he was still getting low grades in all of his classes. During Rod's junior year, when the cost got to be too much for his dad, he transferred back to Tahoma High School.

By now he had been gone for over two years, and

the bond with his old friends was different. A new guy, a skateboarder from California, had one of those personalities that drew other kids to him. Rod fell in with that crowd; a crowd that thought life was carefree. They messed around all the time, did whatever they wanted, and didn't care about the consequences. What they really did was take drugs and consume alcohol, first weed and beer, then acid and hard liquor. He was an exceptional athlete, and the football coaches were excited when he came back to Tahoma. The next fall he turned out for football, practiced with the team for only two weeks, and then quit. He just didn't care anymore about football or anything else. His life was beginning to unravel and he was heading down a different path.

Rod stood staring down at the cars speeding beneath his feet. It was three o'clock in the morning and he was standing on an overpass, freaked out on acid. He no longer knew what was or wasn't real. Mesmerized by the drone of the traffic, and watching the headlights blur by, he thought to himself, "I wonder what it would be like to jump?" A quiet voice *(was it real?)* told him to keep his feet on the ground and don't jump.

The evening had started with a Homecoming party at his old private school. The wild party eventually moved to some nearby hotel rooms, and was fueled by alcohol and a variety of drugs, lots of drugs. Rod knew what to do with them. Suddenly, the cops arrived and busted into the hotel room. While trying to escape, he knocked down an officer, and then ran across the court-yard with the rest of the partygoers. They were finally caught, hiding in a laundry room, but were only told to pack up their things and get out of the hotel. Why the cops didn't arrest them was a mystery.

That night was a distorted fog of taking drugs and partying, scuffling with the cops and hiding in the laundry room. Standing on the overpass had really freaked him out.

When Rod noticed a girl named Sacha, who had moved to Maple Valley from Bellevue, Washington, his friends said she was too straight for him. She was a church-going girl and attended Maple Valley Presbyterian Church. He was still taken by surprise at how pretty she was. Rod and Sacha dated a little. On three or four occasions they had even gone to her youth group meetings. Her church had built a skateboard ramp, so he went there often with his friends, just to skate and hang out. Even then, Rod was usually high on drugs. Weece Masterson, a firecracker of a Brazilian lady, was one of the volunteers from the church and led a short Bible study. She could always tell when they were messing around and high on drugs, and would call them on it. Sacha had asked Rod to quit using all drugs, but he wouldn't listen. He felt he could and should be able to do whatever he wanted.

One day Rod sent Sacha a poem. When she discovered that a buddy of his actually wrote it for his own girlfriend, she was furious. She confronted Rod and broke up with him. This was fine with him. He didn't think they were that serious anyway, and it wasn't a big deal to him. But a small, quiet voice inside seemed to be pulling him back to her. One night, out of the blue, he did something that was totally out of character for him. He wrote a letter to Sacha. It was the closest thing to a love letter he had ever written, and she was deeply touched by his words. Not long afterwards, they started going out again.

Her youth group sponsored a snowboarding trip to White Pass. During the outing, Weece talked to Rod about letting God into his life. "Let go of the wheel and let God do the driving," she said. This was good advice, but Rod wasn't ready to listen.

A few weeks later, Rod stayed with a guy who was his ride to go snowboarding the next morning. He had a real "trippy" room that was all set up for hallucinogens. They had been doing acid all evening and Rod felt like he was sinking into quicksand and suffocating. He was afraid to shut his eyes, fearing he would never wake up. Rod knew that acid really messed with his mind, and he tried to be careful who he was with when he got stoned. This guy stood behind him, making distorted faces in a mirror, which made Rod think he was seeing evil faces leering at him. When he screamed, "Knock it off!" the guy denied doing anything. Maybe Rod passed out, he wasn't sure, but eventually he was able to split, desperate and paranoid. He had no memory of leaving the place and driving away.

Rod found himself at Weece's door at 2:30 in the

morning, stoned and freaked out. Weece and her husband saw immediately that he was a mess. They laid him down on the floor and called the paramedics. Because of the drugs he had taken, his heart was racing out of control with a heart rate of over 200 beats per minute (an average, healthy heart rate is in the 60's). He felt as if he was suffocating and dying. Everything went black. The next day, he regained consciousness in a hospital room, hooked up to heart monitors and intravenous tubes. Since he was eighteen, it wasn't necessary to tell his parents what had happened. They were a close family, and when he finally revealed to his dad that he had overdosed on acid, his dad cried. He had no idea that this was actually the second time Rod had looked death in the face because of drugs. Rod became frightened that the drugs would eventually kill him. He decided he would quit, cold turkey. It wasn't going to be that simple.

After the church snowboarding trip, Rod and Sacha started dating more seriously. In February of her junior year, Sacha found out she was pregnant. They talked to Rod's mom because they thought she would be most able to handle the news. Offering them her support, she said, "We'll work it out." They knew that telling Sacha's mom would not be easy. They waited to tell her until they thought she might notice Sacha's condition. Her pained disbelief was hard to take, but not a surprise to either of them. Morgan was born that October; Rod and Sacha were married the following March.

Working at his dad's pizza parlor made it easy for Rod to get lots of beer and stay partied-up. But he knew it was time to grow up and be responsible. After all, he was a husband and a father. He asked himself the question, "What am I going to do with my life?"

His dad's friend told Rod about a technical college he attended in Phoenix, Arizona, the Universal Technical Institute. While there, he received training in heating, ventilation and air conditioning, and was now making some pretty good money in that field. Rod decided this was what he wanted to do. He asked his dad if he could use some of the money from the large settlement he was awarded as a result of his eye accident. His dad, who was guardian of the funds until Rod was twenty-eight, was skeptical about giving him any of it because of his bad track record. He finally relented, and agreed to release some funds to help him with his education.

The young family moved to Phoenix, looking for a new start. It was great to be out on their own, just the three of them, with their own place. Fifteen months later, Rod graduated in the top two percent of his class. While they were there, he stayed clean and sober, using no drugs whatsoever.

After graduation, they moved back to Washington. He got a good job working in an HVAC company, affording them their own apartment. Another child, Kaleb, was born. Life was good. But it didn't last. After one year, the company was sold, leaving Rod without a job. That short window of time had only been the calm in the eye of the hurricane. A cycle began of more jobs, and more moves to Montana, Phoenix and Seattle. More than once, during the next several years, the companies

Rod worked for were sold, and he was unemployed again.

The young family took a lot of financial hits, each one taking another bite out of Rod's settlement money. While they were in Phoenix, Rod saw an opportunity to start his own business, and paired up with a friend from the technical institute, who was also one of the top students. They developed a business plan, and to get things going, Rod used his money to finance the new venture. This partner was a single guy with no family responsibilities, who enjoyed partying and smoking dope. Because the business didn't generate enough work to keep both guys busy, Rod easily yielded to old temptations, and slipped into a laidback lifestyle of getting stoned, golfing, smoking and playing cards. When he wasn't at home some evenings, he used the business as an alibi for Sacha. It soon became apparent, however, that his partner was using only Rod's money to purchase everything, trucks, tools and materials, as though it were monopoly money. An argument ensued, and he quickly pulled out of the business.

Telling Sacha that the business was folding and they were moving back to Maple Valley was not easy. They had lived in Phoenix for three years. She had made some very close friends, and was active in Bible studies. Her mom had also transferred to Arizona to be near her daughter and the grandkids. Their third child, Hunter, had been born there. Rod was relentless in his decision to leave. He packed everything up, and moved his family back to Maple Valley.

The pattern didn't change once they returned to Maple Valley. Rod was hired at a couple of jobs, but was soon laid off when they ended. It seemed at every new work opportunity he would be paired with someone who was into drugs, even crack and cocaine. He started using drugs heavily again, and it got to the point that he was smoking weed just to get through the day. Rod thought he was concealing it from Sacha by telling her he was working late. With the help of Visine drops, he was able to remove some of the red from his eyes, and smoking cigarettes would cover the smell of the weed. Their fourth child, Molly, was born, and Sacha was kept busy with her kids. They all lived in a large house, with his dad and two brothers, which made it much easier to ignore things that were happening around them.

Rod was miserable at home and at work. He felt the extreme weight and responsibility of having a large family at such a young age. It seemed the only thing he was good at was goofing off and getting stoned. Life was just "blah." When his boss told him they had to lay him off, it didn't even matter. Rod started a pressure washing business, but it didn't fly. The trust fund was almost gone. What was the point of trying anymore? Random thoughts of suicide even hit him a few times. He was lost, confused and feeling trapped. He had run the gamut and was at the end of his rope. He just didn't care anymore.

A few years later, when Rod was thinking about

what he had done with his life, an intense moment of clarity came to him from out of the blue. He reflected, "I'm twenty-eight years old. I've got a wife and four kids. I'm an adult, but what have I done with my life? Absolutely nothing! I know, God, You should be in the driver's seat, so let's see what YOU can do! With my free will, I choose to do YOUR will!"

He didn't tell anyone about his decision; it was between God and him, but things changed instantly. He felt differently about himself, as though he were a new person inside. When he read the Bible, things suddenly registered. He began studying in the book of Proverbs, and thought to himself, "This is good stuff. I wish I had read it when I was younger! All that trouble! This is like a handbook for life." He started listening to that still, small voice that he had been ignoring for ten years. He also quit drugs that day.

Trouble hurried right over to tempt him. Sacha's brother-in-law came to his house to deliver a bag of weed that Rod had ordered six weeks earlier. The weed looked like it was worth three times the asking price. It was too hard to resist, knowing he could make some good money if he sold it. When the brother-in-law suggested they all smoke a bowl to test the quality, he did. While he showered to cover up the smell, the guilt settled hard on Rod. Later that night, Sacha's sister called Rod, demanding payment, but didn't follow their secret system of communication. She blabbed it on the answering machine, which was louder than it had ever been before, and Sacha heard everything. She was furious and hurt, and Rod could not deny any of it.

While sitting on the couch that night, Rod knew it was his fault. There was no one to blame but himself. He started praying, "Lord, I'm so sorry for what I did. If only I can go in there and make up with Sacha, then I'll know You truly exist. I promise I'll go to church and faithfully read my Bible." He really didn't think there was much hope of her forgiving him. However, when he went in to apologize, he was surprised to find Sacha reading her Bible.

They were able to talk, and a lot of feelings came out. Sacha told him, "All of these years I sat by quietly and watched, but I was always praying for you." It really hit him hard that while he was out at night partying and getting high, she had been at home praying. She never demanded anything, and now Rod knew she had been patiently waiting for God to answer her prayers.

"Thank you, God!" Rod knew God had heard his prayer and that He was really there. Rod kept his commitments to God and started going to church with Sacha.

A month later, seven churches in the Maple Valley area hosted a "Purpose Driven Life" campaign. Throughout the community billboards, flyers and ads asked the question, "Got Purpose?" It was perfect timing for Rod. As he read the book, "The Purpose Driven Life" by Rick Warren, he knew God really did have a plan and a purpose for him. Things were clicking.

Joey, son of his friend, Weece, encouraged Rod to go with him to a Bible Study Fellowship. It was there, while reading the Scriptures, that the Bible suddenly spoke to him in a powerful way. This was nothing like the old days when it seemed impossible to decipher. It made sense, and he could now understand it. The Word of God burned in his heart, and he kept reading.

Rod and his younger brother had always been close. Ever since they were kids, they had hung out together smoking, drinking and contemplating the meaning of life. They had talked about the "quest for truth" for years. His brother was now an actor and model in New York. One day, Rod called him saying, "I've found the truth! The truth is really out there and I've found it!"

"What," he said, "are you talking about?"

Rod told him, "Hey, you don't have to hear it just from me. Pick up a Bible yourself and look. You'll find it!"

Their summer-long debate proved how much his brother now believed in his New Age philosophy. He kept telling Rod, "God exists and God is in us, so we are like God." Rod still sent him a Bible, but wondered if he would ever open it. He never stopped praying for him.

Rod and Sacha were still living with his dad, who was hounding him really hard about this "God" stuff. He saw the changes happening in Rod, and didn't like how religious he was becoming. Previously, Rod had cussed like a sailor, smoked two packs of cigarettes a day and chewed tobacco for years. These habits all disappeared, and instead of golfing on Sundays, he went to church. His dad constantly hammered him with every possible objection why he should be helping him instead of going to church.

A missionary was speaking at their church, talking about a trip to Nicaragua with an organization called Agros. The men would work in a village, installing an irrigation system to help the people become self-sufficient. Sacha knew Rod had an interest. She encouraged him to sign up, telling him, "You know you really want to do it!" Rod always liked helping people, and this would be an opportunity for him to give something to someone else.

His dad was very upset that Rod wanted to go on this mission and said, "You've got no job, no money and four kids to feed. There's no way you can do this. You've got so many responsibilities, but if you do go, you shouldn't have to pay." Rod knew there was some truth to what his dad was saying and decided to call Don, the team leader, to cancel his reservation before the plane tickets were purchased. Just as he was reaching for the phone, it rang. It was Don!

"Hey, Rod," he said. "The church has decided to pick up the tab for your trip. You won't need to pay for anything unless you want to take some spending money." What amazing timing! Rod knew this was God's doing. God was just getting warmed up. The next seven months were a roller coaster ride packed with amazing events that built Rod's faith in the Lord.

While he was visiting his mom in Montana, telling her about God, he got a call from Sacha's mom. She knew Rod was looking for a job and had heard Costco was hiring a maintenance technician. When he got back in town, he went to Costco to apply. On the way, he felt he should contact an old childhood friend, a buddy named Dan. He wanted to let him know about the changes God had made in his life. They hadn't seen each other for at least ten years, and back then, he considered Dan a fanatical Baptist. At Costco, he filled out the application and went to the service desk to ask for a paperclip. The lady working behind the counter recognized him, saying, "Rod, don't you remember me? I'm Laurie, Dan's sister-in-law." She looked over Rod's resume and said, "You know, this is what my husband does. You should apply where he works, they pay a lot more. I'll call him for you."

Rod applied for the job, a really good one, and got it! He is still working there today with other Christian men, who are helping him to build his faith in God. What are the chances that the very person he felt led to contact would help him get an interview that day? Amazing coincidence!

At a meeting about the upcoming trip to Nicaragua, a speaker presented a slide show on what to expect. He showed the picture of a lady who had moved from the East Coast to help the people that were living in the dump of the capital city. With excitement, he said, "This is Ruby, the original 4x4. We call her that because she is four feet tall and four feet wide. If you get a chance to meet her, definitely do it!"

Rod felt strongly that Ruby had a message for him. The feeling continued, so he prayed, "God, let Ruby know I'm coming. Let me meet Ruby."

It was November, 2004, when the Agros volunteers stepped off the bus into the Nicaraguan jungle. The sensory overload was intense. Howler monkeys screamed deep, booming shrieks from the high jungle canopy, and parrots squawked noisily as they flapped onto the tops of the shorter banana trees. Even the geckos chirped like hidden birds in the bushes. The humidity, the tropical sun beating down on their heads, and the noise in the seething, shouting jungle was overwhelming. Only the soft fragrance of the jungle flowers was gentle.

Oh, and the people! Rod wondered how the people could have such happy smiles when they lived so poorly. Their homes were just shelters pieced together out of scraps of metal, tires, logs or whatever else they could find to tie around the trees. They didn't have a decent source of water. They could barely feed themselves. Their clothes were ragged hand-me-downs. Yet, they were the happiest-looking people Rod had ever seen. In spite of their poverty, they were eager to feed the team of fourteen men and four women. They cooked corn to present a friendly meal to their guests.

The Agros men were digging out a canal and building an irrigation system. The women worked with the kids. Because the program's goal is self-sufficiency for the villages, there is a strict policy that you don't take much with you. Everything is built with tools to which the villagers have access. The Agros organization buys land for a village, helps them get set up, and then sets up a workable buyback plan so the villagers can have long-term independence. The volunteers were instructed that no one was to bring gifts or handouts, but should come only with willing hands and Christian love. Rod brought all of it, plus a humble and grateful heart. He couldn't believe he was actually there.

The first day was tough. The tropical sun baked the men as they cleaned out the mud-clogged canal. By lunchtime, everyone was ready to get out of the sun. They had packed their coolers with cold drinks and sand-wiches. Hot and tired, the team crowded into the bus to eat, ignoring the village kids and families outside. The kids were hungry, too, and asked for food. They were told there wasn't enough food to share. Because it was against the volunteer contract to give handouts, they were turned away. It felt very wrong to Rod, so he and a few others pulled sandwiches out of the coolers and gave them to the kids anyway. Since Rod and the others had broken the rules, a big debate took place around the campfire that night. Many were upset that they had given sandwiches to the kids; others were glad they had. Emo-tions ran high with some women even crying. Rod sim-ply couldn't believe anyone would refuse to feed hungry kids.

Even though he was usually quiet, Rod spoke out very passionately. He told them, "If God calls me to help these people then I'm going to do it. If I refused and God asked why, I don't want to be without an answer." He spoke for a long time, stirring the people with strong words that had a very deep impact on them. They were surprised that a quiet guy like Rod could have such a burning and powerful conviction. Rod knew the words he had spoken came from God, and was amazed that God was using *him* to speak His message. Since Rod didn't normally speak out in public, people really noticed when he did. God received all the credit.

It was not on the itinerary to travel to the capital city, but on Saturday night, the leader asked, "Does anyone want to go meet Ruby?"

Rod and a few others spoke up, and said, "Sure!" He still knew in his heart he was supposed to meet her, but didn't think he would get the chance. The leader described her as a flamboyant, Pentecostal prophet. This made some of the people very skeptical, and they weren't sure they believed in those things.

Rod was last to get off the bus, and when Ruby saw him, she gave him a piercing look that burned right through him. She cried out, "You're the one! You're the one! I've had dreams and visions about you for months!" Those were the same months Rod had been praying to meet her! No one else except God knew about those prayers. It was the icing on the cake!

After going into her house, she asked if anyone wanted to be prayed for. Rod knew she had a message for him, and said, "Yes."

As Ruby began praying, anointing him with oil, she mentioned an argument he had with a "young man." He knew immediately that the "young man" was his younger brother. As she prayed for his brother, Rod instantly saw a vision of him and his other family members in hell, in overwhelming darkness and emptiness. In agonizing grief, he broke down and cried, as he never had before. She said, "It's not up to you to save him. Be patient and let God work."

One of Rod's constant prayers was for God to keep him on His path, and not let him again fall into the wicked ways of this world. Ruby continued praying over Rod saying, "God knows you don't want to fall off His path." Rod's head was spinning. Was this really happening? God spoke powerfully through Ruby that night, and Rod had no doubt whatsoever that God had planned their meeting.

After she had finished praying, the others asked him, "Did she hit on anything?"

"Yes!" he cried, explaining that the young man she mentioned was his brother and that they had been debating about God all summer.

One evening, an Agros team leader read a story written by C.S. Lewis about a boy and a lion, and things that continually happened to the boy. The story told how the lion was there all the time and made everything happen for a reason. This lion is an analogy to Jesus, who is always with us. The firelight and jungle noises added to the drama of the reading.

"A long, snarling roar, utterly savage, ripped into the night. The horse and his boy swerved around and galloped as hard as they could away from the beast. Just as they stopped, trembling with exhaustion, another roar exploded again, this time from their left. A distant horse

screamed as her rider's back was raked by the cat's claws. Two more roars, one on each side, drove the horses closer to each other until they were racing side by side in the black night. Thus, the lost boy met the brave girl who would travel with him, changing everything. Their journey would lead to his forgotten homeland, to the King, who was his father, the one who knew his true name. Finally they came face to face with the terrible, beautiful Lion that had shaped their destinies. After one look at the Lion's face, they slipped out of their saddles and knelt at his feet, speechless with wonder.

'I was the Lion. I was the Lion who forced you to join the princess. I was the Cat who comforted you among the houses of the dead. I drove the jackals from you while you slept. I was the Lion who gave the horses the new strength of fear so that you would reach the King in time. I was the Lion you do not remember who pushed you to safety when you lay near death, to where a man sat, wakeful at midnight, to receive you.'

'Then it was you who wounded [the princess]?'

'It was I.'

'Who are you?' the boy asked.

'Myself,' said the voice, very deep and low so that the earth shook. And again, 'Myself,' loud and clear and gay. And then the third time, 'Myself,' whispered so softly that you could hardly hear it, yet it came from all around you as if the leaves rustled with it.

The boy was no longer afraid that the Lion would devour him. A new and different sort of trembling came *over him. Yet, he felt glad too."* (Paraphrased and adapted from *The Chronicles of Narnia, Book 3, "The Horse and His Boy"* by C.S. Lewis)

As Rod sat by the campfire listening to the story, he knew he was like that boy. His journey had started with a horrible roar, blinding him with pain the day the knife hit his eye. Death had drawn close more than once. He couldn't see it then, but the hand of God was protecting him throughout his journey. It took sixteen years, traveling down a rocky road, before he yielded to God. On the way, he was joined by a brave girl who traveled with him, patiently praying. Finally, their journey of troubles drove him to the One who knows his name. Now they kneel together at His feet, thankful to see that God was there all the time.

Epilogue

Rod and Sacha's story is not finished. It has just begun. If you talk to them, you will hear two humble, grateful people who saw God take a life that was completely broken and make it completely new. Rod is quick to share what he has learned so far.

✧   Before, when I heard people say "Born-again Christian," it had a bad ring in my head, like a hillbilly in coveralls, or some wild TV preacher. I know differently now. I am new from the inside out. God reached in and changed me.

✧ It's been a two-year time of studying and learning to listen to God. I'm learning to be obedient.

✧ I want each of you to know that God has a plan for your life. He knows your name and wants a relationship with you. I can feel His presence all the time.

✧ When I look back at my life and at all the things that happened to me, I can see how God was protecting me. God was there all the time.

10

I Was Pronounced Dead

The Story of Steve Bowen
Written by Melanie Rosecrans

Beep … Beep … Beep. That was all I could focus on, and it was the first sound I had heard in nine days. I tried to lift my arms to push myself upright, but they were strapped down to two cold, metal bars. Over the loud beeping noise in my ear, I could hear voices in the distance, but I couldn't seem to open my eyes. I lay there wondering what was going on. Was I dreaming? As I focused harder, trying to open my eyes, they slowly cracked open. Brightness swarmed in, causing me to squint with pain. All I could see was intense, bright, white light. I slowly turned my head and saw more white, white walls, white ceiling, white curtains and a stiff, white sheet covering my body. Turning my head back, I tried to focus on a dark figure that was standing in front of me. It was a woman I didn't recognize, and she was staring at me with a look of bewilderment on her face. Where was I? How in the world did I get here? I slowly opened my mouth to speak, and with a soft, hoarse voice whispered, "I want a Bible and I want to go to church." I had never uttered those words before.

Over the next few hours, the mystery of where I was and how I got there was solemnly revealed. I was at Harborview Hospital, and had just come out of a coma that I had been in for nine days. The puzzlement I saw on the woman's face was because the doctors didn't expect me to live. While I was in the coma, the doctors had taken x-rays, which revealed a hairline fracture to my spine that could paralyze me completely. If I *did* come out of the coma, I would probably never walk, talk or even function normally. In fact, at one time, I actually flat lined long enough to be pronounced dead. This would explain the bewildered look on the face of the woman standing beside my bed when I woke up. This woman, Olivia, was my wife, and she told the story of my accident.

I had gone to work on January 28, 1991, like any normal day. My assignment was to work on a traffic signal up in the bucket truck. Brian was directing traffic below. A little while later, he left his flagging position to quickly put his tools in the service truck. During his absence, a semi truck-trailer came along, and with no one to direct traffic, plowed into the bucket truck. The 300-pound bucket broke off the boom and fell 25 feet to the ground. I was strapped inside.

Brian came running over and began CPR on me immediately. He continued with this gruesome task until someone was able to relieve him, freeing him to call 911. When the paramedics arrived, they couldn't believe someone was actually performing CPR on my bloody, mangled body. They said they wouldn't have even tried when they saw the horrific damage that had been done to my entire head. If Brian hadn't attempted to keep me alive, there was a great chance I would never have gone into a coma. I would probably have died before arriving at the hospital. I had sustained massive head injuries, a fractured jaw, a shattered right knee, and a severed left optic nerve, resulting in permanent blindness to my left eye. While at the hospital, doctors had to remove one of my ribs to help reconstruct my distorted face.

I was at Harborview for many days, and when out of danger, was transferred to Northwest Hospital. After being released from Northwest, Olivia and I moved in with her parents in Seattle, so they could help with my care. I was finally able to do what I wanted when I first came out of my coma: go to church and read a Bible. I had never been a religious man before my accident; church and God were all very new to me. I don't think I was sure why I needed to go. I only knew I had to go to church to get through this tragic time in my life.

Due to life-long disabilities and permanent damage to my eye, I received a sizable settlement. A portion of it was invested, and we were also able to purchase a home in Moxee, along with new vehicles, toys and animals. I assumed these possessions would bring some sense of fulfillment to my life. I was wrong.

Things were a lot different for me because of the trauma caused to my head. Thoughts didn't come as quickly now as they had before the accident. My movements were not as steady, and my physical strength was not as robust. One day flowed into the next, and I was content not doing much of anything. I couldn't work or continue with all the hobbies and activities that I loved to do. I just wasn't the same person. Gloom was my companion, and often getting out of bed was too much of a challenge. My life and my thoughts were headed to a dark, dark place. Thoughts of suicide dominated my mind. As I drifted from day to day, I tried to figure out my reason for living.

At first, I was dependent upon Olivia to handle the money transactions and to take care of all our paperwork. When I was able to start answering the phone and to check the mail, I discovered we were in a financial mess! Creditors were calling daily, and the mail was flooded with overdue statements and collection notices. Olivia had not paid any bills for months, and had used most of our investment money for her own pleasures. Amongst all this, she asked me for a divorce. Not only was the money pretty much gone, but I also had to suffer through a long, drawn-out divorce that took two and a half years and thousands of dollars before it became final.

In the midst of all this, God placed a miracle in my life. He sent one lone creature that would help save me from my depressing existence. Her name was Rozie. She was a three-month-old, white Samoyed/German Shepherd mix who wandered into my life, abandoned and in need of care. Although she was malnourished, mangy and full of worms, I took her in and nursed her back to health. Everyone thought she was ugly and ragged, but all I saw was a beautiful creature who just needed my love. We immediately bonded and became partners. Wherever we went, we were side by side. I now had something that made me happy and made life more meaningful. I felt like living again. She needed me and I certainly needed her. God used Rozie to help bring me out of my darkness and depression.

I've wondered since then…is that how I appeared to God before I accepted Him as my Savior? Mangy, ragged and ugly? In the process of healing, inside and out, I have learned that God did not see me that way. He saw in me exactly what I had seen in Rozie, the beauty of the soul inside. As I have gotten to know God better, I've developed a relationship with Him. He is my partner and we walk side by side. I can trust Him to love me and bring me through anything that life throws my way, just like I was able to nurse Rozie back to full health by loving and caring for her.

After my divorce, I was left with very few material possessions. I moved to South Cle Elum and lived in my pickup truck canopy at my parents' home. I had only my clothes, a sleeping bag and, of course, Rozie. I was eventually able to buy a Cover-It to place over my truck as a tent, but it wasn't what you'd call luxurious. That was how I lived for years, saving money until I could finally buy my own home.

Since the accident, things hadn't been that great for me. Still, I continued to go to church in South Cle Elum, and knew that God was touching my life in a positive way. I wasn't angry about any of the things that had happened to me, and I forgave my wife and others who had hurt me. I knew that God had a better plan for me. He wouldn't have kept me alive after the accident if He didn't have a purpose for me on this earth. I began to recognize all the hidden blessings God was placing in my life. I was alive! That in itself was the biggest blessing of all.

My life was finally coming together, filled with a measure of joy and contentment. However, my best friend, Rozie, had become quite old and weary. It was heartbreaking for me to watch her slow down and age. She ultimately became so weak, so ill, that I knew I had to do something to release her from her suffering. Sadly, I decided to take her to the veterinarian to be euthanized. It was one of the most difficult decisions I ever had to make. I felt horrible as I loaded her up in my truck. I got behind the wheel and began driving to the vet's office, wondering if I was making the right choice. Just before I arrived at the office, I looked over at her and noticed something was different. I pulled my truck over to the side of the road and leaned towards her. My best friend was gone; she died peacefully on that short ride to the vet. The Lord, once again, had proven Himself to be loving and full of grace. Rozie had fulfilled her mission by giving me a purpose for living, and providing me with complete joy when nothing else in the world could. God took her out of my life just as perfectly and significantly as our initial encounter on that first day.

My transformation from lost soul to steadfast Christian continued, with the help of the people at a church in South Cle Elum. They soon became my friends, and more importantly, my family. We walked together with Christ, attended Bible studies and services, and helped guide each other through our daily struggles. I returned my gratitude to my church family by helping out with the maintenance and care of the Lord's house. I plowed the church parking lot during the winter, repaired broken appliances and fixtures, picked up and delivered the mail, and provided any handyman services they needed. I still enjoy doing these things for the church today.

Families in the church embraced me, and it was through them that I realized what my mission was in life. It was to bring happiness and joy to others, and to testify to them of the greatness of the Lord. What better person to do this than someone who has had a real testimony of God's grace in his life? I enjoy helping people and giving as much of myself as I can. That is the mission God has given me. He knows our strengths, and He helps us use them so that others can experience Him too. I feel blessed to give and bring people joy because, in turn, it is the one thing that brings me great happiness. I still have to face many trials in my life, but I've learned to take it all in stride. I always have a smile on my face because I love the Lord. I thank God for the gift of life He has given me.

Having never thought about God before my accident, I now realize He has been with me all my life. He has always wanted me to know Him and to be His disciple. God used my accident as a way of getting my attention, so I would listen as He spoke to me about His amazing love.

11

Something Was Missing

Written by Marlene Drew

My mind was made up. I was leaving the jerk, my husband of seven years. Nothing could stop me–not another death threat, not another beating, not even locking me up and holding me hostage, all done in the name of love.

It has been said a woman will play out many scenarios in her mind before she takes action. She plays the "what if" game. I believe it because that is what I did. In my mind, I was constantly rehearsing the various scenarios, the benefits and the consequences. What if I leave, how *would* I survive? What if I leave, how *would* I do it? What if I stay, how *could* I survive? Those questions kept haunting my mind. This time I was really going; I was gone, G-O-N-E! I was leaving with my daughter, and had a sense of peace and relief at having finally made the decision. Hastily packing our few belongings and picking up my daughter from school, I headed to the Greyhound Bus Station for the long trek from Washington to my sister's home in California. Somewhere in this drama was a lesson to be learned. As we "left the driving to Greyhound" across a few states, I had plenty of time to think about what I should do. My story will reveal that it was a very long journey before I discovered the true answer for my life.

I grew up in the Seattle area, the baby of the family, with good, solid parents. Typical of that generation, there was never a lot of affection in our home. My dad was hard, strict, and at times, extremely critical. He ruled the roost with the proverbial "iron fist" and "velvet glove," though the velvet wasn't always so soft. We didn't go to church, but my mom and dad had a defined moral code, one I seemed to constantly be at odds with. I really can't say why, but I was rebellious and more than a little on the wild side. I attended high school during the "hippie" era, and flaunted the rules of convention along with the best of them. I started drinking when I was about fifteen or sixteen, and also tried some of the harder drugs, acid a couple of times and some speed. I didn't care for them. Drinking and smoking pot suited my lifestyle better. Sex was something I did without much thought. It could have had something to do with the fact that I drank so much at the parties that my judgment was severely impaired. All my friends were doing the same thing. But hey, wasn't high school for partying, sex, drugs and rock and roll? That's exactly what I did.

I still had a feeling that something was missing, that there was more out there, but I had no clue what it might be. A girlfriend even took me to church with her a couple of times, but that didn't seem to give me any answers.

For excitement and adventure, I ran away with my boyfriend when I was only seventeen. We headed to sunny California with no plan; we were just going to live off the land. The border patrol caught us as we tried to get into Mexico, and turned us into the juvenile authorities. My parents were contacted, and I was told to come home. I did, and even finished high school, but the drinking, smoking and drugging didn't stop. It provided the thrill that I craved and didn't know how else to find.

After graduation, I moved to the University District where there was always something happening; parties were everywhere. I was nineteen years old and a full-blown alcoholic, drinking every day and continually smoking pot. Hard liquor and marijuana colored my every decision. That's where I met my future husband. He was cute and liked to party, which worked for me. Was there supposed to be more? Nobody said anything to me (or maybe I wasn't listening) about looking "deeper" into the person; about getting to know them BEFORE you decided that you're in love, and about getting married BEFORE sex. Now that would have been just ridiculous. I mean, how are you supposed to know if you are in love if you don't have sex? Aren't they the same thing? I sleep with you; therefore, I love you! Who knew? Obviously not me! Three months later, I was pregnant with our daughter. We decided to move to California, and some time after his divorce was final, we got married.

My husband drank as much as I did, and matched my pot habit with his own. We were poor because my prince among men did not like to work, and I only worked at the local Arctic Circle. He did, however, like to keep tabs on my every move in case I should take a notion to try and leave our ever-increasing abusive situation. I wrote a letter to the mother of an old boyfriend just to see how things were going up north. My husband found it and was enraged. That was the first time he hit me. He clobbered me in the ear so hard that it bled. On another occasion, he took me for a ride up in the hills that surrounded our town. I had committed some offense that irritated him, and he wanted me to know that next time, he would bring me up these winding roads and drive the car over the edge. I don't even remember what I did to provoke him. All this time, I kept working, kept drinking and kept up my pot smoking.

For some reason, in the dead of winter, he decided we would move to Spokane to be near his family. I remember it well because he took all my shoes, except for my sandals, and threw them away. Why did he do that? Two feet of snow was on the ground, and he thought it would keep me from going anywhere.

Our daughter was born during all of this chaos. After having her, some maternal instincts were awakened within me, but I can't say it brought out anything paternal in my Romeo. The strain of having another human being vie for my attention was very difficult for him, and it began to show in his behavior towards me. He got more abusive, physically and verbally, and continually told me how stupid I was. I was reminded, daily, that I would never be anything worthwhile to anyone. I was even required to walk ten feet behind him. When he threatened to kill me if I tried to leave, I finally got the clue that the honeymoon was over. He wouldn't allow me to get my driver's license because he thought the lack of a license would slow down my ability for flight. During all this, I was working, raising a daughter, juggling my efforts to keep my abusive husband happy, and still partying. I had stopped using hard drugs (I didn't consider pot a drug), but I was still drinking. So here we were, in Spokane, Washington, living among his family and friends, and picking up life right where it had left off. We just partied with a new group of people. I would simply drag my daughter along if anybody mentioned the word "party."

When I was about twenty-four, I got my driver's license and did exactly what my husband thought I would do, which was leave. He had always made threatening promises about what would happen to me if I were to ever move out. That scared me, but I knew I had to get my daughter out of this abusive life. I called my folks, told them we were coming home, and headed out across the state to Seattle. The next day, my husband called to see if, or when, I was coming home. I told him I wasn't going back. He screamed at me over the phone, "Just look outside!" He had come all the way over from Spokane to take the car. Doing what I know many victims of abuse do, I agreed to meet with him and he convinced me to come back home. At that point, I couldn't have told you why I did it, but now I know that abuse victims get caught up in the fear and the lies the abuser tells them. They make you think that no one else would ever put up with you because you are so unlovable, and they are doing you such a huge favor to let you stay with them. Somewhere along the way, you actually start to believe it. Finally, you feel sorry for *them* because of how horrible you are, and for hurting their feelings. Go figure! By now, I had been an active alcoholic for almost ten years, and it wasn't easy for me to discern what was real anymore.

I did go back with him, but, mentally, I was already making plans to leave again. The deal breaker was when my daughter saw her dad getting ready to go out and asked, "Are you going to get drunk again?" Talk about "out of the mouths of babes." She was only six, and knew more about what was right and wrong than I did. Upon reflection, it has struck me as odd that something as simple as the words of an innocent child would be the catalyst that would spur me into action.

That is what prompted me to call my dad to ask if we could live with them for a while. This time he said, "No, I feel it would be better for you to go further away, where you would both be safe." He sent us money to take the bus to my sister's home in California.

My sister and her husband, a civil engineer, lived in the Sierra Nevada hills in a beautiful home. They took us in willingly, and were such a great help. I didn't drink very often during that time, even though I still had the urge. I was able to keep it under control. We spent about six months there, living in all that peace and quiet. A part of my sister's life was attending church and I went with her a few times, but I still didn't get the meaning of it at all. My attendance was more for appearance sake than any actual relationship with God.

While living with my sister, I missed the excitement and the adrenaline rush I had always craved in my life. Those missing elements caused me to listen to my husband when he called to tell me how much he missed me. I broke down and said, "Okay, we'll come back."

When my sister heard him declaring his devotion to me over the phone, and then heard me agree to go back to him, she grabbed the phone and shouted, "Absolutely not!" She didn't understand that I just didn't want to be alone, or that I really missed the excitement. The lecture she gave me brought me back to my senses, and I finally snapped out of my momentary loss of sanity. We stayed at my sister's home, choosing not to return to the chaotic mess from which we had just escaped. Looking back now, it's pretty sick to think living with an abuser is exciting, always wondering if each day might be your last.

I knew I needed to go back to Washington and get a divorce. When my parents came to California for a visit, riding back with them was the easiest way to make this happen. Once again, I packed up our things and headed back to Seattle to live with them. Life was pretty bleak; I was broke, facing an imminent divorce, I had no husband, no job and a little girl who needed me. My parents were such a big help in every way, and encouraged me to take some positive steps. I started going to a business college, applying for a program through welfare to obtain some financial support. I was still struggling with the drinking on the weekends, but would have moments of clarity during the week. I had not been this sober for years. Going to school awakened some knowledge in me that I did have a brain, and more than a fair amount of intelligence.

An old boyfriend, whom I had seriously dated in high school, passed through town. We clicked and began to go out with each other again. In high school, he wanted to go to college, which he did. I wanted to party, which I did. Our relationship finally disintegrated. Now, years later, we started seeing each other again, and had a great time. However, it was difficult to be together since we lived in different states.

One day, I hit bottom, realizing what a mess I had made of my life, and went down to the beach near my parents' house to walk by the water and cry it out. I found myself praying to God to help me, which was weird since I didn't really have a relationship of any kind with Him. Once again, that same feeling came over me that there was more out there, that something was missing in my life. I had always connected that "missing" feeling with a man, hoping that if the right one loved me enough, it would fill the void.

On my way home, I checked my mail. I had received a letter, along with a marriage proposal, from my old, now new, boyfriend! I knew this must be the answer to my prayers. I accepted his offer, we moved in together, and started making plans to get married. Here I was again, turning to a man to find that missing something. It wasn't long, however, before I realized our engagement was a mistake, and I broke it off. The divorce from my prior husband also became final. The only thing he had wanted was the gun I had previously taken to keep him from shooting me. The gun was returned, and that was that! Two relationships ended.

I finished school, landing a job as a secretary with a local real estate agency. I still kept my drinking to the weekends only, and I considered it under control. I guess you could have called me a "weekend warrior." I have to admit, while in an alcoholic haze during those binges, I didn't have much discretion. I didn't care which guy I went home with, which guy I brought home, or even about getting the name of some of the men I was intimate with.

A great-looking guy bought a house through the agency where I worked, and I started dating him. He was handsome, a charming, Italian, smooth talking kind of handsome. It appeared he had a "good" job and was buying a house, what more could I want? He did not drink much, that was my department, but he was a very heavy pot smoker. I was shocked when I discovered that I was pregnant again. This should have been a wonderful time, a great feeling, knowing that I was carrying a life inside of me. There was one little drawback to all of that: I didn't know if it was my boyfriend's child or someone else's. I was *fairly* certain it was the child of my current love interest, but I couldn't say that with total conviction.

I thought about my life, where I was and where I was going. I wasn't married and didn't relish the thought of getting locked into another disastrous liaison. I made the decision to not have the baby. It seemed to be a smart choice, and it certainly seemed to be a popular, socially acceptable, method of birth control. I could name half a dozen women that had already had abortions, and it certainly didn't seem to be bothering them. If society, in general, says abortion is an all right thing to do, then who cares? The ads say a woman has the right to choose; so I chose, for fifty dollars, to take care of my little problem. I hadn't told anyone about having the abortion, including my boyfriend, and went alone to have it done, not that he would have cared. I was somewhat certain the baby was his, but he was pretty sure it wasn't. After it was over, I felt a guilty pall come over me. What was this all about? I shut off that emotion as quickly as I could by having a little drink to fix these feelings of wrongdoing, setting it all aside and pushing on, by golly.

My previous pattern was emerging, and I married this charming Italian. I seemed to be the only one who didn't get it. My, oh my, did I pick a winner this time! He kept losing these "good" jobs, so I had to keep on working, despite my drinking. I wonder if losing his jobs had anything to do with all the negative traits that I hadn't seen before: his violent temper, his unstable mental condition, the blackouts he would have and his ability to quickly flip into a rage. I knew of all those things first-hand, including his drug abuse. What I didn't know was that he was sexually abusing my daughter. I was clueless at the time, but in looking back, the signs were there. He spent a lot of time at home "between" jobs, and when I came home from work, my little girl was always outside playing, in all kinds of weather. She never told me what was wrong, and I didn't ask. He would threaten suicide or homicide at any talk of a breakup, and I knew if I left, it would be "over my dead body." I worked, self-medicated with alcohol, and kept up the cash flow as best

I could. There were big mortgage payments to be made on a nice house, which had to count for something. So why was I so empty inside, and why did that same funny, little feeling keep coming back to me at the oddest times, the one that kept me thinking there was something more for me, that something was missing?

The police had been called one night when my husband's anger erupted out of control, and he had to spend the night in a state-run mental facility. He was a very bright man, a master manipulator, and managed to convince the doctors that this outburst was just a one-time thing. He told them it was stress related from problems at home caused by a nagging wife. He was released on the condition that he would go to counseling with me. I'm certain he felt the counselor would go along with him, and wholeheartedly agree that I was the problem. Otherwise, he would never have been willing to do anything. Somehow, we ended up with a Christian counselor. I know now that this choice of a counselor was no accident; it was God's divine intervention. We went for counseling, both separately and together. The longer we were in counseling, the more obvious it was to the counselor that this marriage was over. Once again, I ended yet another disastrous, destructive relationship. I kept the house and didn't move out, but that also meant I had to make the big mortgage payments. I worked really hard to keep up, but I also continued drinking, which made me feel okay about everything.

Of my own free will, I went back to the counselor. I remember sitting in the office talking about my life, all of it: what had happened during high school, all my broken relationships, my drinking days, right up through to the abortion. I told him of the guilt I felt from having an abortion, and that I knew it hadn't been the right thing to do. But when the shame got too strong, I would just stuff the feelings down or drown them in alcohol. The counselor began telling me about Jesus, and how He would forgive me. I didn't believe it. Of course, I had heard of Jesus; I would have had to live in a vacuum not to have heard something about Him. In my mind, He was more like an ordinary person, and the people I knew rarely forgave even little things. How in the world could He forgive me for all the sinful things I had done? Even though I wasn't sure about everything I was being told, there was a certain amount of curiosity about this Savior. Was there really someone who could wipe away my sins, someone who could love me unconditionally? That was a concept I couldn't grasp. I had worked hard at being loved, and the results weren't very pretty. I think it was the first time I had heard the truth, and it stuck with me. A seed was planted, and with those words, a little glimmer of light shed upon the darkness of my life. I now had the faintest bit of hope, but no idea what to do with it.

One night, driving home from a counseling session, I felt the same old feeling that something more was out there, that something was still missing. This time it wasn't just a twinge; it was more like a flutter on my shoulders, a palpable feeling. The words the counselor had spoken were whirling around inside my head. I started to cry and prayed, "God, please, please help me!"

I just wanted Him to take the weight off of me, to take all my sins, all my mistakes, and the guilt of the abortion. I couldn't carry the load anymore. I couldn't go on this way, and I was crying out for forgiveness. He reached out to me and put His hands on my shoulders. I felt them there. It was the gentle touch of a loving Father, the healing touch of a loving God that I had craved. I had always looked to the touch of a man to heal my hurts. I never thought that the most powerful, loving emotion I had ever felt would come from someone I couldn't see. Our loving Father knows what will touch us. He knows what will impact our lives in a non-refundable way. My Savior knew that my love language was touch and He used that to reach me. I heard His voice, not audibly, but in my heart. He took my tears away and lifted all the sins of the past from me. I let His love come into my heart in place of sorrow. I had such a feeling of peace that nothing in my life could compare to what I felt at that moment, no drug, no alcohol, nothing. The experience was so poignant I have no idea how I got home. Angels must have been with me. I have always regretted that I didn't go back to that counselor to say "thank you" for the gift that was given to me: those words that Jesus would forgive me.

I would like to tell you that I stopped drinking right away, but that would be a lie. I can tell you, with assurance, I now knew who Jesus was, and I knew He was real. People around me began to see the changes, even though I didn't tell anyone about what had happened. A friend of mine even gave me a Bible as a gift. I began going to church regularly and on the day of my baptism, my parents came to watch. My brother gave me a birthday card on that day, symbolic of my new life. I wonder if he knew more than he was telling. The changes in me weren't huge or immediate, but they kept on coming steadily, as I attended church, learning more about my Lord. I felt my faith growing, but sometimes I didn't know what to do next. I was having ups and downs, but I knew that my path was going to go forward in the Lord, no matter how small the steps might seem sometimes.

I met my current husband about twenty-four years ago. He was not a church-going person, had no faith in God, and we were both drinking. I was trying to temper my alcohol use by then. He was using alcohol to dull the pain of a nasty divorce, and being separated from his two little girls. I saw things in him I liked: kindness and a very tender heart. The Lord knew this was something new and different compared to how I usually judged men. We became true friends, not lovers who struggle to care about each other when the sexual attraction is over. At first, we didn't attend church together, but I would always go. Not once did I ever think of giving up my new faith. Because of our children and our past experiences, we didn't rush into marriage. We dated for five years, and when we did marry, I knew I loved him, and he knew he loved me. This wasn't a marriage of convenience, but one ordained by God.

Soon after we were married, we started attending church together, and he also realized there really was something more to life. Things did not happen right away for us. We accepted change slowly. As time went on, God, in His great wisdom, continued blessing us with more and more understanding and love, love for Him and love for each other.

It wasn't a marriage "made in heaven." We both carried a great deal of baggage, mostly brought on by our drinking. There came a point, ten years later, that my husband did something stupid and hurtful, and it devastated me. For the first time in my life, I asked God for guidance. I put my relationship with my husband in His hands; I didn't leave. I thought about it, struggled with it, and finally decided to wait on God. The answer He gave me was, "I forgave you, and you need to forgive him." When my husband came to me, he laid everything at my feet and told me the absolute truth. He asked for my forgiveness and I felt so loved. His deep abiding love for me made me really listen and forgive him. Oh yes, I was hurt deeply, and it wasn't just whooshed away in a stream of tears. I realized that I was seeing the genuine article and I chose to stay. I would not have remained in my marriage if God had not been important in my life, if only a man had met my needs, and if God had not forgiven me. I would have once again been the loser. God has blessed this union and it is wonderful. The odd thing is when my husband destroyed my faith in him, he came and told me what he had done, and God gave me the strength to listen and forgive. Our marriage is stronger than ever. We talk about everything now. Neither of us drinks anymore. We attend Family Life, and participate in marriage-strengthening Bible studies. We both are so very aware that, indeed, there is "something more" out there!

Looking back, I marvel at the way God works; how He can bring about events and circumstances that shape the direction of our lives that cause us to follow Him. If you try to analyze things according to the laws of man, it makes no sense. When I met my husband, we were both drinking. Taking that into consideration, along with my history in relationships, you could only think, "Here she goes again. Doesn't she ever learn?" But God knew what I needed and what my husband needed as well. He knows the character that lies in each one of us, and sees the worthiness that we don't recognize in ourselves.

With God by our side, reminding us of His forgiveness, we have made a commitment to each other to work through our problems. I am so thankful that our faith in the Lord has become the anchor to our marriage. It is more than I ever dreamed would be available to us.

12

A Father to the Fatherless

The Story of Letha Ihrke

Written by Melanie Rosecrans

The cool breeze whistled through the trees in our yard that unforgettable October night. The branches creaked and moaned as they danced powerfully in the wind. I could picture the rope swing on the weeping willow tree, swaying back and forth in the fierce gusts. The glowing moon was creeping into the sky, adding some light to my room, which was filled with the darkness of the night. I was seven years old, and I was still scared of the dark. I was terrified of being by myself in the eerie darkness. Almost every night for a long time now, I had been crawling into bed with my mom and dad down the hall to ease my fear. Dad was starting to put his foot down about it, and thought it was time I learn to sleep in my own bed. Just last night when I was frightened again, I walked down the hallway to their room, peeked in, and tiptoed to my dad's side of the bed, whispering, "Scoot over, Daddy." I knew he didn't want me doing this anymore, but he just opened his eyes, smiled his loving smile, and lifted up the covers so I could crawl in beside him.

Even though I was still afraid of the dark, on one particular night I stayed in my room, falling asleep in my own bed. My slumber didn't last long because a few hours later I was awakened by a frenzy of activity down the hall. My mom kept walking past my bedroom door, going back and forth to my two older brothers' bedroom. They were twelve and seventeen years old. I could hear her crying each time she passed. There were also strange voices and noises coming from my parents' room– unfamiliar voices, full of panic and concern. The only other thing I could hear through the fear and sadness in everyone's voices was some sort of machine making air noises. It sounded like loud breathing noises, in and out, in and out. I sensed that everyone was very upset, and something had happened.

As scared and unsure as I was about what was happening, I still couldn't get out of bed; I was too frightened. It was almost as if there were an invisible hand holding me down, keeping me right where I was. I lay with my covers pulled up to my chin, and listened to the ongoing chaos. My face was flushed and my breathing short; I could hear my own heart pumping hard and fast. Something dreadful had happened. I knew it was my dad. I didn't understand why mom didn't come into my room to tell me what happened. "What was wrong with my daddy?" I could only lie in my bed listening, terrified at what I didn't know.

It seemed as though I had been by myself forever, when our pastor's wife, Eva, came into my room to be with me. I'm sure my mom thought I was asleep, and decided to not wake me during all the chaos. She was trying to cope with the situation as best she could. I was grateful that Eva came in to be with me because I loved her. She sat down on my bed and just held me. She didn't tell me what had happened or what was going on, but I knew. I knew deep in my heart that daddy had died. I cried and cried, and she cried with me. She hugged me, prayed with me, and just comforted me, giving me much needed love and affection.

The next morning was like an awful dream. I only found out for sure that my dad had really died when I overheard my mom talking on the phone with a relative. He'd had a heart attack right down the hall from me. When he went to bed that night, he told my mom, "My ticker's not working right." His chest was hurting, and soon after that, he had his first and only heart attack. He was aware of what was happening and told my mom, "Don't let Pete (my nickname) see me like this." My mom called the ambulance, but when it arrived, all they could do was try to revive him. He died at the young age of thirty-nine; my mom was only thirty-six.

Aunts, uncles and cousins from near and far began arriving at our house. There was also a large outpouring of love and concern from our small community and church. I remember mom had to take me shopping because I needed a new dress to say goodbye to daddy. The family had to go to the "slumber room" for viewing his body the day before the funeral services. The first thing I saw when I walked into that room was his casket, right there in front of me. I had never seen one before. I slowly walked up to it and looked in. As I stared at him, all I could think was, "He's right here; he's just sleeping." I knew that he had died, but the denial was so strong, that the truth was hard to bear. It was just too much for my little mind to understand. I have no memory of what happened next, but I was told later that I became hysterical and lost it emotionally. I'm sure I was in shock, and I almost went berserk right there in the funeral home.

Before all this, I had a pretty amazing and perfect life. My family was very close, figuratively and literally. My grandparents lived about two blocks away, and many of my aunts, uncles and cousins lived in the vicinity as well. Our home was in Kittitas, Washington, a very small town of only five or six hundred people. We had a large home with a big yard, a creek running through it, and a weeping willow tree with a rope swing. We searched for crawdads in the streams, and drove around on our tractor, always having fun. We had bunnies, chickens and other animals running around. We felt safe and secure in that neighborhood, never locking our doors. Without any concerns, I was able to walk to any family member's house when I wanted to visit. Our house seemed to be the meeting place for family and friends. Dad would pull out his guitar, and we would have our own built-in entertainment. He was very special to us and was totally involved in every aspect of our lives. It wasn't unusual for him to get on the floor to play and wrestle with us. One of my favorite memories is when he would come home from work with candy in his shirt pocket, just waiting for me to find it. It really was quite the life!

Everything was different now. Dad was gone, and mom wasn't the same. Her life was filled with sorrow, and she didn't know how to deal with it. She had lost the love of her life, and we certainly didn't want to cause her any more grief. Mom didn't talk about dad's death, so we didn't speak about it either. This silence was a huge elephant in the house that we just couldn't seem to deal with. We all thought we were coping, but looking back now, I know we never grieved properly. We didn't share our feelings of sorrow and loneliness; we just functioned on a day-to-day level.

Since my dad was gone, the security of our home and community was no longer there for me. I was even more terrified of the dark at night, and didn't feel safe in my own house. I remember mom wanting me to sleep with her in her room. One night, someone tried to break in the garage after we had gone to bed. From then on, mom slept with a crowbar beside the bed. I guess she thought she could use that as a weapon if someone ever tried to get into our house. That frightful experience only seemed to worsen my fears.

The next summer came quite quickly, with the fall and winter passing in a blur. My oldest brother left home and joined the Navy, which left just three of us. Everyone had been very supportive and kind to me following my dad's death. The pastor's daughter, Judy, was my best friend, and she, along with her mother, Eva, played a huge part in getting me through my pain and sorrow during those difficult months. I became very close to the pastor's family, including their other children. Eva doted on me, and would always greet me with a huge hug. When Judy and I were in the second grade, a very special thing happened to us. Our mothers decided it would be fun to be room mothers together for our class. Eva was an amazing person and I loved her dearly. At that time in my life, I never dreamed that one day I would grow up to be a pastor's wife, just like her, that I would play the piano for church, just like her, or that I would love and teach children in church, just like her. I had such a good example, and it always stuck with me. I wanted to be just like her.

I regularly attended Children's Church, which was held in the pastor's home next to the church. This was a familiar place to me; it was my best friend's home, and I felt very secure there. Eva would share stories about our Heavenly Father and His love for me. When she asked if anyone wanted Jesus to come into his or her heart, I raised my hand to say yes. I remember that morning well. We sang, "He Owns the Cattle on a Thousand Hills," which told me of the greatness of God. God met me there and I never forgot that moment.

The following summer, however, a second tragedy occurred in my life. Eva was involved in a near-fatal accident, and was immediately transferred to a hospital in Ellensburg. She was in very critical condition. My mother went to pick up Judy and her younger brother to bring them to our house, while Pastor Mark went to stay with his wife. The remaining children were sent to other family members in the church. Judy and her brother stayed overnight at our house. The next morning, my mother pulled me aside and told me that Eva had died during the night. I was heartbroken. We thought Pastor

Mark would get his children that morning, but he was unable to do so until later that day. We couldn't tell Judy and her brother what had happened; they needed to hear it from their father. Judy and I played together all day long, and it was awful for me knowing that her mother had died and I couldn't say anything to her. I had to keep my emotions in check, pretending everything would be fine and trying not to fall apart. I had already experienced the sorrow of losing a parent, and I knew the pain and grief Judy would soon face. My heart ached for her. I also ached for my loss of Eva, this amazing woman, who had such an impact on my life. Another important person had been taken from me, creating a gaping hole in my young heart. It was a very long day in my life, and I will never forget it.

After her mother died, Judy and I didn't get to see very much of each other. Her family's relatives were constantly at their house to help with the care of the children. Being only nine years old, it was a very difficult and lonely time for both of us. Pastor Mark and his family eventually relocated to another town. My very special friend, and her family that I loved so dearly, had moved away. As I look back, I can see that God placed these people in my life at just the right time to guide and nurture me, as a father would.

Soon after Judy left, a new pastor, Dan, and his wife, Sandy, came to our church. They also touched my heart and left a lasting mark. They took me under their wings, and treated me like a member of their family, even taking me on vacations with them. Pastor Dan was like a second father to me. In fact, I felt so close to him, I asked if I could call him "Dan" instead of "pastor." He laughed and said I should probably just keep calling him "pastor." They were at our church for only three or four years, and then moved to Ellensburg. They didn't just walk out of my life though. They were close enough that I could call whenever I needed some fatherly advice. During those formative years, they shared with me the message that God would carry me through every situation in my life that I would be asked to endure. A seed was planted that our Heavenly Father would always be by my side and always in my life, like my father.

One evening, when I was in eighth grade, I knew God was truly walking with me. Mom was working the late shift that night, and I decided to go over to my oldest brother's house. He was back from the military, had recently married, and was now living in Kittitas. It was dark outside, but my brother's house was only a few blocks away. As I walked along, a car pulled up next to me, and the man inside asked if I wanted a ride. Being from a small town, I didn't have a clue about the dangers a stranger might present. I just told him, "No, thank you, my brother's house is right up the road." For some reason, I remembered the license plate number on the car as the man drove away. I told my brother what had happened. He immediately called the police, giving them the license plate number that I had noted in my head. We soon found out that the man was a convicted child molester. I couldn't believe it; trouble in my small town? I just knew that it was God, my loving Father, who had protected me.

That next year, my younger brother left home and joined the military. When he was sent to Vietnam, I was terrified, thinking he would go to war and never return. I didn't think I could handle the loss of someone else that I loved. But God helped me to endure the death of yet another loved one, my grandfather. It was "grandpa" who did all those things with me that dad would have done, had he lived. I loved him dearly, and missed all the times we had together. My freshman year in high school was a very long year. I was waiting and praying for my brother to come home safely from the Army, and also dealing with the loss of my grandfather. I remember consciously making a decision to never be close to anyone again; it was too painful to love them and then lose them. Oh yes, I still knew the Lord was there and that I could always call on Him, but the pain of continually losing people I loved was too heartbreaking for me. Still, God, in His great mercy, has always kept His hand on my life, showing me over and over that He is my Heavenly Father, and is always with me.

Throughout high school, I remained close to the Lord. When a troubling time would come up, He was the one I would cling to, and He was my stability. I believe our family still silently wrestled with the thought, "If He is such a loving God, why would He take away our father like He did? Why do bad things happen to good people?" There had always been an underlying belief in my mind that you should not ask God "why," but should accept your life and deal with it.

I was able to put these questions into perspective when I learned that Christ, while on the cross, cried out, *"My God, my God, why have You forsaken me?"* I knew that if Jesus could ask "why," then so could I. I also learned that you could just keep asking God "why" until the "why" just doesn't matter anymore. I always felt fortunate that I had God by my side; that He was the one who brought me through all my difficult days.

I was very blessed by knowing the new pastor of our church and his family, but our relationship had a little different twist. They didn't just influence my life with teaching and support. They actually set my future life in motion by introducing me to the man who would later become my husband. Here is how it happened.

The pastor's wife had two brothers in the Army. A girl in our youth group was writing to one of them, and without my knowledge, told him all about me. She wrote of our adventures together, sharing some of my personality with him. When he came home from the military, we finally met. And that was it! He was the man for me, and the man I would eventually marry. We went to Northwest College together, but I had to quit because of a sledding accident. He continued on with more education.

After we got married, we lived in South Seattle, and had three children. Our love for the Lord then led us to Enumclaw, Washington, where we served as youth pastors. Sumner, Washington, was our next home, and we attended a church in that community for eleven years. Later, we had the great privilege of pastoring a church in the Cle Elum area, and through adoption, we added a fourth child to our family. Life has been wonderful and very fulfilling to me, as a wife to my husband and mom to my four children.

When I was just thirty, another earth-shaking experience in my life happened, which was a huge turning point. After the loss of my father, the doctors had prescribed Valium for my mother to help treat her sorrow and depression. Valium was the "wonder drug" back then. She had been taking it for so long, it began having a negative effect on her body and she became very ill. When her doctor recommended she attend a 12-Step meeting to help her with this problem, I decided to go along with her. During these meetings, I pondered our home and family, and realized how dysfunctional our lives had been. It became clear how I had been an enabler to my mother, by allowing her to ignore the sorrow in our lives, as well as her own. Since we had never had a discussion about dad's death, we were never able to share our grief. Later that night, after going home from the 12-Step meeting, I lay in bed crying uncontrollably. It was then I realized that from the time I was that little seven-year-old girl until now, I had never mourned the loss of my father. That night, as an adult, I was finally able to grieve. I am so grateful for the doctor who helped us see our needs and recommended the 12-Step Program. God was with my mom and me when He guided us to that meeting.

I was finally able to leave the pain and troubles of my youth behind me, and move on in life as an adult. It was my turning point, my dividing line. I Corinthians

13:11 was the scripture verse that I thought of that night as I let my childhood go: *"When I was a child, I talked like a child, I thought like a child, I reasoned like a child. When I became a man [an adult], I put childish ways behind me."* God had healed me, and I was finally able to move forward in my life without the baggage from those painful childhood years. I came to the emotionally healthy decision that my dad, also, would have wanted me to move on.

My mother and I remained very close. She developed Parkinson's disease a few years ago, and in mid-October of last year, her body began to shut down. Since my father died in the month of October, it had always been a difficult month for her. My brothers and our families were always by her side in those last days, but we still wrestled with the question, "Why did mom have to suffer?" The big "why" question. In spite of that lingering question, I still felt like I was emotionally able to handle her death. This time, I was able to deal with it in the right way, and it was a very healing time for me.

Throughout my life, God has always been by my side. There were many people taken from me that I loved, but He continued to guide me and love me. Each time I had to face pain and hurt, it made me a better, stronger person. When I wasn't dealing well with situations in my life, God would place a person or an opportunity in my path to guide me. He placed the right people in my life when I needed them, protecting me through all my experiences. Psalm 68:5-6 is a passage that rings near and dear to my heart: *"A father to the fatherless, a defender of wid-*

ows, is God in his holy dwelling." He has been my Father for my entire life. He satisfies the void in my heart left by the loss of my real dad. When I picture my Heavenly Father, I am crawling up on His lap and He wraps His arms around me, just as my natural father did so many times when I was young. God has always provided unconditional love, protection, security, and anything and everything a young girl's heart would need.

I feel blessed that I was able to have such a wonderful, loving, natural father. Even though I lost him when I was very young, I knew when my dad died that he loved me. There was never any question about the love he had for all his children. It breaks my heart when I see kids today dealing with their dad's leaving, not coming home, or not paying child support. I don't know how they handle the fact that their dad isn't involved in their life. When people refer to "God the Father," the normal thing to do is reflect upon your natural father and his love. For children who have never felt the love of their real dad, how can they even relate to the love from a Heavenly Father? I see it as the work of Satan to rob innocent children of a father's love, and ultimately hinder them from having that loving relationship with their Heavenly Father. It seems unfair that I lost my dad at such an early age, but it was a blessing that I got to experience real fatherly love while he was alive. I had a gift that many never get to experience, and I feel very fortunate for that. It was truly a gift from God!

On our twenty-fifth wedding anniversary, my husband and I went to visit our son in Washington D.C., where he was stationed in the military. While there, God impressed upon me to put some of my life's trials into words. I don't know if I understood why, or how they came to me so easily, but I could feel God's guiding hand once again, helping me to heal all my pain and sorrow. I wrote a poem describing the loss of my father and the effect it had on my life:

THE OLE' WILLOW TREE

As a child, in my front yard,
a weeping willow grew.
A stately tree with drooping leaves;
as I grew, it grew too.

I learned to climb in the ole' willow,
the first time I recall,
I was most brave about going up;
coming down I was afraid I'd fall.

I remember daddy would come to the rescue;
I always knew he would.
I'd go a little higher and he'd come get me down,
we both just understood.

As I grew older with confidence strong,

Coal to Diamonds

I'd climb to the highest of limbs.
From that very top perch,
I could see all the houses
of relatives, neighbors and friends.

The willow was secure and serene,
a place for my God and me.
My God and me, and the willow tree,
and my daddy, what sweet memories!

A dreadful day came when God took daddy;
such deep pain I recall.
Who would come to my rescue?
Who would catch me now, if I were to fall?

The willow and I, we cried together.
I identified with its weeping.
I took comfort from its strong curved limbs;
'twas there I found safekeeping.

My God and me and the willow tree,
there I'd muse the questions of life.
"Why did You take daddy?" I would ask,
"It brought us all such strife!"

'Twas there I learned life was not fair
and you deal with what you're given.
Time goes on and wounds heal,
and daddy wouldn't want me to stop livin'.

A Father to the Fatherless

I'm all grown now with my own family;
I'm still fond of weeping willows.
I still miss my dad every single night
when I lay my head on my pillow.

I'll always be grateful to God above
for growing that tree for me.
For listening to me, and comforting me,
and for wonderful memories.

Letha Ihrke

13

Through It All—We Made It

The Story of Mike and Stevie Haynes
Written by Marlene Drew

Stevie blinked as she walked through the door of the courthouse. It was a beautiful, bright day, Indian summer in the Northwest. There was a light breeze, and the smell of fall was prevalent. It was the best time of the year to live there. She didn't notice any of these things because her life, as she knew it, had just ended. The date was September 29, 1994, her tenth wedding anniversary, and she had just filed the papers to end her marriage to Mike. Trying unsuccessfully to stem the flow of tears, she walked across the parking lot to her car. Once inside, she sat behind the wheel and let the tears fall as her mind played back over the years.

Stevie met Mike when she was nineteen and he was twenty-two. They both lived in the same Issaquah trailer park, she in a trailer, and Mike in a small apartment complex.

Being a child prodigy of the counterculture, Stevie was no stranger to the recreational drugs that were so readily available in those times. She was a girl on a mission, out to score some pot for the upcoming weekend. A man she knew to be a dealer was sitting on the grass in front of the apartments. He looked up as she approached, and Stevie made her request, something like a takeout order at a fast food drive-thru. The order could not be filled at that moment, but the man knew someone who might know another man. That someone was described as a blond-haired, pony-tailed guy riding a three-wheeled VW trike. Stevie waited around until the guy, whose name was Mike, pulled up on the trike. They met, and she made her drug desire known. When she looked into those blue eyes that twinkled back, Stevie thought she was in love. More than just a transaction for pot was made. She left a little piece of her heart behind.

Mike, on the other hand, was not so easily captivated. His life was consumed by going to the beach, hanging out in bars and riding his trike. Who needed a girlfriend who couldn't share all that? Stevie wasn't old enough to "hit the bars" yet. They traveled separate paths until a chance meeting happened at the Issaquah Salmon Days Festival. Mike offered Stevie a ride on his trike, and that was the beginning of their relationship. It wasn't long before the two were living together. Their mutual passion for living large, having fun, and using drugs and alcohol to exhilarate the experience cemented their relationship. After five years of living one extended party, Mike and Stevie were married on September 29, 1984. They settled down in Renton in a little brick house with cedar trim, a small carport off to one side, a little patch of lawn in the front and back, two bedrooms, a bath, a kitchen and a living room. It was sufficient for a young couple whose sole purpose in life was to have fun, fun and more fun, and at any cost.

Stevie worked as a hairdresser in Issaquah, and

Mike was a mechanic in Fall City. When they both went off to work in the mornings, Stevie would have a joint tucked away in her purse for later, and Mike would take a fifth of Wild Turkey in his toolbox to fortify himself for the drive home. Evenings found Stevie at home smoking pot, her drug of choice. Mike, having partaken of a little libation from the toolbox stash, would stop off at a bar on his way home to top off his tank. Weekends were devoted to partying with friends and lots of booze, pot, coke and meth. Life was a party! Sometimes entire paychecks were consumed by their favorite recreation with no cares about tomorrow.

After a time, the second bedroom, which now served as a growing room for marijuana, had to be cleared out in anticipation of the stork's arrival. Three years later, another child was born. Life went on spinning dizzily out of control, high to high, paycheck to paycheck, until one day…one fight…the last fight!

Stevie had stepped out the front door, inadvertently locking it behind her. In her drug-hazed mind, she was somehow sure that Mike had locked her out. Stevie's anger boiled over and she kicked the door in. Mike, seeing her inexplicable behavior, got equally mad. He picked up an aluminum baseball bat and started beating holes in the wall. After the shouting, screaming and cursing subsided, they were left with a front door that wouldn't close, a dozen holes in the living room walls, and a marriage that was truly in shambles. No one said much right then, but a seed had been planted in Stevie's heart. What had they become? The next day, when her three-year-old son, in a fit of childish anger, picked up the same aluminum bat and started hitting the wall, something really hit Stevie. She called it her "light bulb moment" or her "reality check." She looked at her son and wondered, "What have we done? What are we doing?" Stevie knew a drastic change needed to happen or their lives would just continue on the same old way, and their children would suffer for it.

Stevie loved Mike, but her mother's heart was stronger. She made arrangements that day to move out, saying nothing to Mike. Friends and family, with pickups and cars, cleared out the house and moved Stevie to her sister's home, storing her belongings in the garage. One week later, Stevie found herself crying in the parking lot of the courthouse after ten years of marriage.

When Mike left the bar that night and headed home to his little piece of heaven, the little brick house with cedar trim, a small carport off to one side, a little patch of lawn in the front and back, two bedrooms (one that had served as a growing room), one bath, a kitchen and a living room, he had no idea that every room would be empty. He stood in the doorway and surveyed the carnage the hasty exit had caused. His heart pounding in his ears was the only sound in the room. Nothing had prepared him for this. He was devastated, bewildered and in shock. There was no little wife sitting on the couch getting high, no little girl, no little boy, no note, nothing. Gone! Mike picked up the phone, which was about the only thing left in the house, and called his younger brother. "Help me," he cried. "Stevie's gone! She took the kids. I don't know where she is." His brother told him to just hang on; he and his wife would be right over.

As the couple was driving to Mike's house, they wondered what God had in store. They had been home less than one hour from a vacation that had been cut short by a sense of foreboding. Both of them had been overcome with a feeling that something was wrong. They had no idea what it was, but it was strong enough for them to cut their trip short. The car's engine hadn't even cooled off when Mike called. They wondered, "Was this God at work in Mike and Stevie's life? Was He getting them to finally see the wrong road they were headed down?" They prayed for the wisdom and strength to be able to minister to Mike.

Mike met his brother at the door, where the two hugged each other tightly and cried together. His brother could see that Mike's emotions were reeling. Sitting on the floor, they began to talk to him about the Lord. The couple spoke to Mike about the direction his life had taken, voicing their concerns about the possible consequences to himself and his family. Because of the choices he was making, he needed to consider what affect this lifestyle could have on his young children. He also needed to think about how this, in turn, would affect the choices made by them in their adult lives. Mike's brother explained, "God wants more for you than this. He wants you to put your trust in Him. He has the power to cure you of your alcoholism, heal your marriage, and give your family back to you."

Mike knew he was hearing the truth, and within

half an hour, prayed to accept the Lord into his heart. An amazing thing happened, all his cravings for alcohol and drugs were gone from that moment on! Mike knew he still had a long road to travel, but he headed down it willingly. The very next day, he enrolled himself in counseling for drug and alcohol abuse, as well as anger management and parenting classes. He began attending church regularly with his brother, and prayed daily for another chance at his marriage.

Two months later, with the Lord alive in his heart, Mike headed off to find Stevie. He had a dozen roses, a card in his hand, his heart pinned to his sleeve, and a determination to get his family back. Stevie, meanwhile, had not talked to Mike since she had literally "cleared out" of the house. Money was tight, times were hard and the kids needed to eat. Settling into survival mode, she had decided to have a yard sale with the absconded household goods. Stevie was totally involved in her enterprise when Mike arrived. She was not ready for any contact yet, so Stevie went inside the house. Mike said nothing, just handed the flowers and card to her sister, and then purchased the microwave, a dresser and a chair, all at full price with no dickering involved. A short time later, during one of the perfunctory child exchanges, Mike asked his estranged wife for a date. Stevie was reticent to say the least. She had no intention of embarking down any path that might lead back to the road she had so heartbreakingly exited. Mike persisted, saying, "I thought we could go to an AA meeting." Stevie looked at him, really looked into those eyes she loved so well, and saw something different. She didn't realize that she was seeing the fire of the Lord in Mike, but she knew something had changed, and it was genuine. Ironically, their first date, a real date, began at an AA group meeting many years after their drug deal meeting, and a true, loving courtship followed.

Mike talked to Stevie about the Lord. He was still attending all his classes, and was beginning to build a relationship with his Savior. When he talked, Stevie listened, and something, a yearning, a longing, started to stir in her heart. Could this be the missing element, the lifelong search for a "feel good"? Could it have always been the need for God? Mike took Stevie to his brother and sister-in-law's house, and the four of them spent countless hours talking about God and His great love for them. They began to attend church together. While they were at a baptismal service with the church at Lake Washington, Stevie prayed and accepted Jesus as her Savior. Later that day, Mike and Stevie shared in the wonderful experience of being baptized together.

The couple continued to date, usually going to AA meetings together, and then for coffee or dinner afterward, talking as they hadn't done in years. They talked about their dreams and their disappointments, sharing their goals as a couple, as parents, and now as Christians.

A significant change occurred for them as they attended church and became a part of a church family. The members opened their doors, as well as their hearts, to the new couple. They prayed for, and with them, for their marriage to be healed. The support they received, and the examples they saw of other healthy marriages in the church, was the cornerstone that helped bolster Mike and Stevie to fix their own union. As God's love started to work a healing in their individual hearts, it became clear that He was also healing their broken marriage. They started to love each other again, as a man and wife are to love, with respect and kindness, seeing one another through new eyes filled with God's love. As their faith grew stronger, so did the love they shared. With the Lord as the guiding force, instead of drugs and alcohol, they began a life together once again.

Now, having celebrated their twenty-first wedding anniversary, they have a marriage that is vibrant, fun and alive in the Lord. It is such a testimony to their children, who have witnessed the healing in their parents' marriage, and ultimately in all of their lives.

Mike and Stevie look back with wonder now at where the wrong roads took them, and how they had sailed along, rudderless, with no captain. Many of their friends' marriages have failed, children have been lost in the system, and lives have gone undefined in a sea of drugs and alcohol. They realize they could have been next on the list of doomed marriages. But God had a plan: He called, and Mike and Stevie answered in faith, a faith that has been rewarded in the most beautiful way. They are alive and well in their marriage and thanking God daily for His grace, mercy and forgiveness.

A little anecdote: As their twentieth anniversary approached, Mike and Stevie attended a charity auction. One of the local attorneys had donated three hours of his time to be put up for bid. When that lot number came up, Mike began bidding to the bewilderment of Stevie, who promptly asked him, "What are you doing? We don't need an attorney."

Mike replied, "Our twentieth anniversary is coming up, and I want to be prepared in case you leave me again." The two laughed together, something they couldn't have done just a few years earlier. Stevie reflected on how solid they were in their marriage, and how God had worked such a healing in them. Today, there is understanding, and even joy, at what they have come through together.

14

Our Retreat Center in the Woods

The Story of Susan Sill
Written by Janet Bunch

I became a cop at the age of twenty-three. I was self-confident, self-reliant, full of myself, and loved everything about my job. The idea of ever leaving the police force was not a conscious thought I could entertain in my mind. It took heartache and devastation to make me realize that I was not in control of my life. God had a bigger plan. My life's journey was just beginning. It would soon unfold, as doors would open and then close. My vision would fade and then become clear, and miracle after miracle would take place.

I was brought up as a Christian. My mother had a strong relationship with God and reflected a wonderful example of Christ. I went to a Catholic school and participated in the youth functions there. Like most teenagers, I had my share of troubles, but was fortunate to make it safely through those trying years. Even while going to college, I made sure I attended church, and always believed that Christ was a good friend. He was someone I could turn to in times of trouble. After college, life took a turn in a completely different direction, and I found myself in unfamiliar territory.

I knew it was wrong to have a relationship with a man separated, but not divorced. The attraction I had for this man, and my own self-will, rose above what I knew to be right. From the very beginning of our relationship, God was showing me Tom was not the right one for me. But despite the warnings, I gave in to my own desires and married him after his divorce. My life with him was very difficult and full of many trials. Still, in all of the unrest, God blessed us with a beautiful son. I tried hard to maintain a false sense of rightness, but after two years, Tom and I separated and not as friends.

I had a two-year-old child, was working swing shift as a cop, and my husband had just left me. I was devastated! I cried out to God, "What are Your plans for me now? I don't know what I'm supposed to do. Doesn't the Bible say, '*if God is for me, who would be against me?*' Have I really known Your presence in my life? If so, why is life so fragile and everything falling apart around me?" All these questions needed answers, and I didn't have any. It soon became evident that I, me, myself, and I, was the stumbling block to finding God's purpose for my life.

Shortly after my cry to God for help, I had a dream. My body was transformed into a tiny, little silhouette formed in the fetal position. I was cupped in the palm of a giant and powerful hand, very majestic and strong. I felt secure and reassured. My Heavenly Father had placed me in a safe haven within His presence, and I was surrounded by His love and mercy. I didn't know where I was headed, but I knew if I let God lead the way, He would guide me and hold me up. The very next day, I felt free, like the freedom only a bird can experience when soaring in the heights of a clear, blue sky, and looking down on the earth with absolutely no fear. I knew then that everything was going to be okay.

The road ahead wasn't a smooth one, but things did get easier. I was still a cop with the King County Sheriff's Office. I had a big, beautiful home and yard in Woodinville. I lived in a great neighborhood and had a wonderful church family. There were several bumps in the road that needed ironing out, but for the first time in my life, I was able to turn it all over to my Heavenly Father. The comfort of letting God lead my life brought a reassurance and confidence that was beyond me. Through release of my own self-will, and by allowing God to bless my life beyond my imagination, the desires of my heart would soon be realized.

I met Craig. We were introduced through mutual Christian friends, and he lived within a block of my home. I was convinced that I was never getting married again, and told myself that Craig was not a part of God's plan for me. No way! It wasn't that I was questioning God's authority, I was just back paddling, determined to never be in rough waters again. God began speaking to me, and finally I said to Him, "I don't know what Your plan is or how You'll work this out, but I think I'm supposed to be with this guy." His plan soon became clear and the rough waters were made calm. With His blessings, we were married in 1996. I was surrounded by love and comfort in a new way.

This relationship was totally God-designed and was reassuring in every way. It was as though God was saying, "I want you to live as free of complications as possible." Little did we know what was just around the corner for us.

Time went by quickly those first few years. When you take your everyday, ordinary life and place it before God, embracing all He does for you, He will bless you mightily. We were filled with God's peace and comfort in every aspect of our lives. He blessed us with two more children, and our life was moving in a positive direction.

One day, I had a vision. It was as though God was giving me a picture of the future, but He gave it only to me. I felt He was telling me that we were supposed to have a Christian retreat center where we could be a witness to Christ. It was all I could think about. When...where...how? We had no experience in that area, so my thoughts were reeling. I asked God, "What in the world are You thinking? I'm a cop and Craig is an engineer. We don't know anything about running a retreat center." Then it occurred to me. This was the next step in God's plan for Craig's and my journey together. But I still didn't understand why God revealed the vision to only me.

As Craig and I were coming home from one of our many "honeymoon vacations," we talked about our future plans, what we might do, where we would live, and what kind of business we would venture into. I was again taken aback by the vision God had shown me, but I was afraid to tell Craig. It was two days later when I finally found the courage to approach my husband and share it with him. I was certain he would think I was out of my mind. I explained to him that I had no idea what this retreat was supposed to look like, or when it was all going to come together, but I knew it was God's plan for us, and He would make it happen somehow. Craig didn't quite know how to answer, and all he could muster up was, "Well, okay." It became clear to me, by Craig's response, why God gave the vision to just me. I knew in my heart that God's plan was for us to open a retreat center, and all we needed to do was follow His lead.

The thought of living "in the woods" out in the country was very appealing to both of us. We started looking at property in Eastern Washington, and after a couple of years, located a 100-acre compound on the outskirts of Cle Elum, Washington. I was so excited! I knew this was it! The minute we found this place, God's doors started opening, and His plan was well on its way to completion, at least that's what I thought. But this 100-acre compound turned out to be a three million dollar project, and we certainly didn't have that kind of money! What Craig and I did have was an agreement to always follow God's will, and for now we felt we should pursue the Christian retreat center. We asked God, "You do want us to follow through with this, don't You? So what's next?"

We discussed how we should approach this gigantic adventure with no collateral and no experience. In an attempt to find ways to get financing, we told as many people as we knew. Miracles started happening, one right after another, and everything appeared to be falling into place. We just knew God was going to make this happen. One afternoon, while we were home relaxing on our deck, the phone rang. I answered it, spoke with the caller, and after hanging up, walked back outside in utter amazement. I told Craig, "You won't believe this, but a major Bellevue bank just called. They heard about our retreat project and want to talk about financing. If we get the Small Business Administration involved, they think we can pull this off!"

So, we met with them. Things were looking brighter and brighter for our future; everything was coming together. At the same time, I felt the pressure of the huge sacrifices we would be required to make. I'd have to quit my job as a cop, give up my big, beautiful home and yard that I loved, give up the great schools for our children, and leave our wonderful neighbors and church family. Yet, we continued to pursue having a life in the country, "in the woods."

Craig had been an engineer for quite some time, and in preparation for our move, he started an engineering company out of our home. We did this for two reasons: first, to find out just how to manage an in-home business; and second, to eliminate the need for a nanny, as Craig would be a stay-at-home dad. His company was very small at first, but was definitely moving in the right direction. Craig wasn't able to earn as much as he did when he was working for another company, but it provided adequately for our needs. Our family was, and still is, his top priority. We decided that if we moved to Cle Elum, we could take the business and explore the potential for expansion there. Our vision was still very excit-ing, but reality was starting to set in. Since we didn't know if the retreat center would happen, especially since no papers had been signed, we needed to focus on the "what if" factors. What if we couldn't get financing for the retreat center? What if we weren't able to sell our house? What if this really wasn't God's plan? What if...? What if...? Feeling God's direction, we continued to press on, preparing to make some major changes in our lives.

I loved our home, and selling it was a very hard experience, and emotionally draining. God was way ahead of us. We had our place appraised prior to putting it on the market, and were told that we should do some minor improvements in order to make it more appealing to sell. When we were finished with the changes they suggested, it wasn't our home anymore, and I didn't want to live there. God merely turned the picture around to a different view and helped my heart adjust to letting go. It was another confirmation that we were being held in the palm of God's hand, and we were headed in the right eastwardly direction.

Several realtors suggested we ask top dollar for our home. Much to our surprise, within twenty-four hours of putting it on the market, we received several of-fers for more than our asking price. It quickly sold, and we were told it could close at any time. In anticipation of our move to the Cle Elum area, I quit my job. Just one week before we had to move, it hit us like a brick wall, "We have to find a place to live over there!" The retreat deal hadn't closed yet, so we looked for a place to rent in Cle Elum. Nothing was available. We looked at a new house to buy in South Cle Elum, but the carpets hadn't even been installed, and we needed to take occupancy the following week. We told the seller we wanted the house, and then told him we had to be able to close in a week. He said there wasn't a problem in getting the house finished, but there was no way to complete the paperwork that quickly. Our home in Woodinville had been sold to a mortgage broker. With his help, the documents were ready for our signatures in only five days. The house was ours at the end of the week! Since we expected the financing for the new retreat center to be finalized shortly, we thought surely this would be our "temporary" home for just a couple of months.

God was opening windows and doors all at the same time, and we just couldn't believe this was happening for us. The retreat property seemed perfect! The entire complex had enough buildings on it for us to get started: a beautiful lodge, a house for us to live in, and eight six-person duplexes. But then the reality set in. To break even, we would need to enlarge the facilities to accommodate more people.

We had been living in our new home for just two months, and the retreat deal was soon going to close. We were excited, filled with hope and anticipation. But caught up in the excitement was a strong feeling of hesitation. Those feelings gave way to doubt, and we began to pray, "God, all these doors are opening. It really looks like Your vision is almost here, but if this isn't the direction You want us to go, we need a sign. We aren't sure we will recognize some little subtle one that tells us to halt. If this isn't meant to happen, we need You to give us a giant billboard-type sign and slam the doors tight." Within hours, two things took place to confirm our prayers were answered. First, the Small Business Administration backed out of the deal. Then the bank backed out of the financing. There was no question that God had answered our prayer, especially with these signs of absolute confirmation. So, that ended that. The vision that was once so vivid had faded into the dim fog of a colorless dream. I was so confused and prayed, "Why God? Why did You bring us here? What are we supposed to do now?"

As the days passed, we realized that we could help with many needs right here in our own community. Adults and teenagers began flocking to our home, showing up at our door just to visit. A teenage Christian organization, called "YoungLife," was in need of volunteers to help with their program. Having a love and a special passion for kids, Craig and I decided to invite YoungLife groups into our home. These kids are fed both spiritually and physically, and kept busy with all sorts of activities. Knowing that our home is a safe haven, some are here for dinner and others spend the night. They come from rough homes, as well as from solid Christian families. There are the computer geeks, the athletes, and the stoners, and they all come to YoungLife and hear about Jesus. Our contact with kids goes beyond YoungLife by helping them with their homework and giving rides to sports activities. When I walk into the local high school, I get teary-eyed because I feel so welcomed. What they need is the extended hand of Christ and someone to love them, just as they are, with no strings attached. Craig and I have found ourselves getting exhausted and tired with so much going on, but God reminds us of what He has orchestrated in our lives, and we are grateful to be used by Him!

While driving a group of kids to a YoungLife Camp with music blaring, food fights abounding, and too much activity for one VW van, a verse came to my mind. *"Delight yourself in the Lord and He will give you the desires of your heart."* Until now, I didn't know how much I really loved teenagers. He has given me a delight in my heart that I didn't even know could exist. If God had told me to give up my house, my job and everything else we had in Woodinville to move to Cle Elum and open our house to needy kids, I would have said, "No way!"

The real eye-opener came when a YoungLife leader, who heard me talking about my vision of a retreat, said, "Don't you understand? What God revealed to you is still very real. That 'Retreat Center' is your HOME." She was right, and so was God. The vision had been real all the time; we had just been focusing in the wrong direction.

When we purchased our home in South Cle Elum, it was to be a temporary residence until we could move to the retreat center way out "in the woods." That dream grew fainter when we realized that the YoungLife kids wouldn't be able to find us as easily if we lived that far out. We wanted to always be available for them. We gave up that dream, replacing it with a new "Retreat Center" closer to town. We were able to purchase property "in the woods," and it is directly across from the high school. God helped us to design our future home, and it will be large enough to hold both our immediate and our extended YoungLife families.

When we were in the process of moving from Woodinville, there was no doubt in our minds that God was giving us a lesson in obedience. He was in control, and the retreat was what He wanted us to do. All we needed was the faith to carry us through and to continue to allow God to be in control. I can remember having a lot of fears in the form of questions. "Is the retreat center real? If we get this retreat center, will we still have time for our children? How will we support them?" I can also remember feeling extremely convicted, and remembering that the same God I envisioned once holding a tiny silhouette in the palm of His hand was the same God that would honor our obedience.

Our lives have been ordered, step-by-step, by the Lord. My ex-husband is now a Christian, and we are good friends. After a twenty-one-year career as a cop, I retired. That profession had its challenges, yet provided me with many wonderful opportunities for witnessing. Now, my job and my calling are to work with the young kids of our local communities, and if needed, offer them a "home away from home."

Craig and I took the walk of faith. God has been

in the pilot seat the entire way. He, and only He, guided us to His "Retreat Center in the Woods." He will always lead us with utmost grace and mercy. The One who loves us unconditionally has orchestrated a beautiful and wonderful purpose for our lives. We pray that He will continue to show us the way to teach our children to be independent and confident in Him. We will always trust in the Lord, giving Him honor and allowing Him to have complete control. Where He leads, we will follow.

15

Amber Waves of Grain

The Story of Kearnel and Sena Lund,
early settlers to our community
Written by Bernice Hagerty
Paraphrased by Marlene Drew

"Go West, young man; Go West." Horace
Greeley's cry rang throughout the nation, inspiring the
hardiest of young men to go and claim their future. It
stole the hearts and grabbed hold of the imaginations of
the brave-hearted, and wouldn't let go. Tales of fertile
lands, with abundant timber, enormous mountain peaks,
and lush green valleys, painted vivid pictures that could
not be set aside.

Kearnel Oliver, "K.O.," Lund had the fever. He
grew up hearing the stories and they burned in his mind
like a branding iron. As a youth, he had an unquenchable
thirst for adventure, but faithfully toiled in his father's
fields, until it was the right time to seek his own future.
He left his home in Nebraska, and headed west for Washington, the Evergreen State, the land of abundance. As
he stood at the top of the last ridge leading into the upper
Kittitas Valley, he gaped in wonder at the craggy, snow-
capped peaks of the Stuart Mountain Range. They stood
majestic before a backdrop of deep sky blue. Stretched
out below was a valley of lush green hues. He knew this
was it; he had found his beautiful piece of heaven,

Xanadu. In an instant, a love affair began that lasted a lifetime. From then on, he would claim this peaceful valley to be the most beautiful sight in the world, calling it home.

Kearnel came from a farming family and always knew he would continue in his father's footsteps...but first things first. In order to finance the purchase of a farm of his own, he would need to devote himself to hard labor and save up some cash. He first went to Sequim, Washington, and worked at a dairy. Later, he came back to his beautiful valley, where he worked for the railroad in Kittitas County, all the while, diligently saving his money. Seeing that his dream of a farm of his own was near, K.O. also realized something else. He could toil away all day long, keep his mind occupied, make money, buy a farm and build a house, but what would be the point? He had no one to share it with, and would be alone in the midst of all this beauty.

His mind wandered back to the beauty he had left behind. Visions of a dark-haired, blue-eyed, corn-fed girl would color his thoughts. Her grace and charm belied the strength and ambition that burned in her heart. Her love of life was enchanting; her zeal for God and church unquenchable. Kearnel knew that conquering the west without her would be pointless, empty and lonely. With excitement in his heart, a song on his lips, and a keen eye focused on the future, he retraced the steps he had taken earlier that year. He traveled to Nebraska to claim the lovely Sena Lee as his bride. On September 22, 1909, they married and began the journey back west to their new home.

Sena's enthusiasm never dimmed throughout the years, and she would often tell the story of their journey west. She told of the beautiful autumn, as the scenery passed by the windows of the train carrying her to their new destiny. Her love for her new home grew apparent, as she marveled at the landscape. She and her young husband began their lives together in a tent alongside the Teanaway River, where Kearnel kept his team of horses, and hauled ties for the railroad. Sena loved living in the camp, watching the snow as it piled up around their tent. The interesting people, and the fact that she was expecting her first baby, added much excitement to her daily chores. With the arrival of spring, they were able to purchase a small farm in the Fowler Creek Canyon. There, they started a small dairy business that kept them afloat and provided their living. The family prospered, according to the times, and enjoyed the addition of two more children.

Kearnel began to have reservations about having purchased property so far away from the main highway and the railroad. He realized that when electricity and the telephone finally did reach the area, Fowler Creek would not be first on the list. He wanted more land to farm, and with his family increasing in size, it was imperative that he make more money to take care of their growing needs. In 1916, a farm on Westside Road, near Easton, was purchased. That land is still in the family today, with numerous members living on the farm, and raising their own children on the very spot that K.O. and Sena carved out their home.

At the turn of the century, neighbors gathered to combine their efforts, and in just one week, a house was built and ready for occupancy. During that era, there was no such thing as wiring, floor coverings, painting and plumbing. But after Kearnel and Sena bought the farm and moved into the house, they installed a flush toilet and a big, claw foot tub. This was considered an extravagance at the time, and there were some people in the community that were a little suspect of having an inside toilet. A large picture of the Nebraska homestead hung in the living room, reminding Kearnel how grateful he was that the Lord had led him from the plains of his Nebraska home to the beauty of the Kittitas Valley. The kitchen was the very heart of the home, with the wood-burning cook stove always sending a delicious aroma wafting throughout the house. Along with the daily meals, Sena baked countless loaves of bread, cakes, cookies and pies for her family. Next to the wall was a long, green bench that sat beside the table, easily accommodating all five children and the hired man. A glass-front dish cupboard held the few treasures that had been gleaned along the way, each one holding a sweet memory of the past. There never was a set of solid stairs in the house. The upstairs bedrooms were reached by yanking on a rope that dangled from the ceiling, with the pull-down staircase providing the access. Measured by today's standards, the house would seem simple, even primitive, but in those times it was fairly modern, and more than sufficient for the needs of the growing Lund family.

The life of a crop farmer then, as now, was a desperate struggle against an economy determined to test the willpower and endurance of every man. The forces of nature held many lives in the balance and left much to chance. Much thought and planning went into the planting of the crops. As soon as the snow was off the ground and the fields dried out, a team of horses was put into their harnesses and the land was cultivated with loving care. The first sprouts of the season elicited feelings akin to giving birth–a birth to the future.

K.O.'s farm produced bountiful crops of potatoes one year, only to find there was no market for them. As disheartening as this was, the pioneer spirit that forged this country rose up in K.O. and Sena, with determination and willingness. They worked from dawn to dusk to eke out a living from the land that held their hearts captive. Getting up with the rooster's crow in the morning, and settling down to rest at night well after the rooster had gone to roost, was all too common a pattern in a farm household. Their lives were not easy, not when measured by life today, but it was the life they knew and loved.

Conservation was part and parcel of every move. Near the house, a large vegetable garden was planted. They grew everything that the climate would allow. Overseen by their vigilant mother, the young children who were not strong enough to work the fields, tended to the watering and weeding. This garden was not a luxury, it was survival! A well-managed orchard, with apples, pears and plums, held court off to the side. The chicken house was full of clucking hens, laying eggs in their nest-ing boxes, and there were always a few young roosters around to accommodate the Sunday dinners. Cows provided the necessary milk, sweet cream, butter, buttermilk and cottage cheese, along with the annually kept steer for the family meat. Sena kept a few sheep for the wool that she cleaned, carded and combed into neat little piles of roving, which was spun into yarn. Her spinning was a hobby that she loved, as well as a provision for the family. The wool socks, hats, gloves, scarves and sweaters were dyed bright colors and were a great source of warmth and comfort for the whole family. The pigs, being an efficient way to get rid of any table scraps left from dinner, were in return, the source of smoked hams, bacon, sausages and chops. Hog fat was rendered for lard, and when the fat was mixed with ash, it became soap to be used by the family throughout the year. Best of all, lard was used to make crusts for the sweetest fruit pies and dough for those delicious, mouth-watering, homemade doughnuts. The root cellar was dug into a dirt bank with a heavy wooden door for access. It held a treasure of goods necessary to sustain life on the farm. The temperature was kept cool and constant by the earth itself. The dirt walls and floors were swept smooth from use. Shelves lined the walls, and up above, wooden rafters stretched from side to side. From these hung the smoked hams, slabs of bacon and sausages that dangled in long, ropey loops. The shelves were organized and neatly arranged with canned goods: fruits, vegetables, meats, jars of jams and wonderful homemade pickles. Barrels of sauerkraut and wooden bins of apples sat off to one side. Carrots and potatoes were packed in straw inside of crates, to help hold them through the cold winter. This seemingly benign little room held the key to surviving throughout the year. It was the virtual lifeblood of the family, housing the results of a year of hard labor.

The year of 1920 brought great hope to the young Lund family. That year's Farmers Almanac forecasted that wheat would be the profitable crop of the season. K.O. set aside the fields with the richest soil for his planting endeavor, and hoped to be able to put the family on easy street, following a good harvest. The fields were prepared, the soil turned, and the seeds planted. The weather held fine with plenty of moisture in the spring, followed by warm summer sun with temperate nights. The green sprouts showed themselves right on time to follow the harvesting schedule. It was quite evident, watching the sprouts develop, that almost every seed had germinated as the little shoots were reaching for the sun. The crop was coming in thick and green. As the stalks grew taller and began to form the heads that would cradle the grain, everyone could tell that something different was happening. The heads were extraordinarily large, bursting with kernels of wheat.

K.O. and his neighbors had never seen anything like it. Word soon spread through the grapevine, as it does in a small community, and people began to arrive from all over the valley to witness the marvelous miracle that was taking place right before them. They would stand back and watch those *amber waves of grain* swaying in the breeze, and talk in reverent whispers about the phenomenon in Kearnel's fields. The County Farm Agent made the trip from the lower valley, which was by no means a small feat in those times, just to see if the rumors were true. He walked the fields solemnly, trying to assimilate what he was seeing, and wondering what had brought on such an abundant crop, as if understanding would be the harbinger for years yet to come. After having examined every plot of wheat, he turned to K.O. and shouted in excitement, "This is the best crop I have ever seen!" K.O., upon hearing the declaration, finally believed it was real.

As harvest time drew closer, the dreaming that had taken place only in their minds began to look as if it would become a reality. After chores were done for the day, and the supper things washed and put in their places, the mail-order catalogs would come out. Little faces scrubbed shiny clean, ready for bed in their flannel pajamas and gowns, would belly down, side by side in front of the fire, pouring over the Sears and Montgomery Ward catalogs. They were looking through the thick tomes that held pictures of things that, up to now, had just been dreams. There were dresses to decide upon, coats and pants for the freezing winters, patent leather shoes, and boots to keep their feet warm and dry. They dreamed of new clothes for school, store-bought, something they had never worn before. The dolls and bright, shiny bicycles that had long been wished for were now, at last, within reach. These decisions could not be made lightly. There were so many choices, and they needed to choose just the right thing. The excitement and anticipation was almost too much to bear. After the kids went to bed, K.O. and Sena would sit together by lantern light, and work on the list of their needs for the house and farm. Sena had been waiting patiently for new curtains and some sorely-needed furniture, and was hoping they would even be able to get those pretty dishes in the window of the General Merchandise Store. K.O. needed implements, new machinery, harnesses and other tools to make life just a little easier for him on the farm.

A banner crop is a thrill not understood by one who has never wrestled with the soil. The work preparing the ground, the planting and watering, while watching and waiting until harvest, all involved long, backbreaking hours. K.O. toiled months on end from sunup to sundown getting the fields ready. Kearnel knew the joy of harvest, but had never experienced anything to compare with the bountiful crop of this year. After harvest, he would usually take the wheat to the granary and wait to see if there was a market for that year's growth. This year, after hearing of his amazing crop, neighbors were ordering wheat from him before harvest time, reserving seed for their next year's planting. They hoped these seeds would reproduce in their fields the same marvel it had done in Kearnel's. Marketing the crop this year would not pose any problem.

The day finally arrived to reap the wheat! The team was harnessed up, driven to the field, and the mower hitched to the tugs. All was in place. The golden stalks fell away from the blade of the mower in neat rows, leaving corduroy stubble across the field, bringing promises of better days. Everyone worked the fields with feverish excitement. Anticipation kept weariness at bay. They knew their efforts would prove worthwhile when their dreams were realized.

Weather, though always unpredictable, could usually be counted on to hold fair during the harvest time. But this year, the skies grew dark and a storm loomed on the horizon, foreshadowing doom for the wheat crop. The prayers began to drift heavenward, "Oh Lord, please don't let it rain! Not on this crop! Not on our greatest crop of all." But the desperate prayers were not to be answered, at least not in the way they had hoped. Nothing could hold back the force of nature when it unleashed the rain, and rain it did. It rained and it rained, and then it rained some more! With broken hearts and heavily-laden spirits, they endured the unceasing drenching from above. Standing at the window, looking at the sodden mess before him, K.O. felt the ache in his heart, and his spirits drooped lower than the remaining uncut stalks of wheat.

When the rain stopped long enough for them to go out, they would thresh some of the wheat, put it in gunny-sacks, and haul the sacks into the living room of the house. The sacks were dumped out on the floor in the center of the room, hoping that the grain could be dried out, and they would be able to salvage part of their dreams. Hot fires were kept burning day and night, and the grain was turned over with a large spade and stirred with a stick to keep the air moving through it. Their house was literally turned into a kiln, with tons of grain piled in every empty spot. Try as they might to keep the kernels contained, the constant stirring and shuffling caused the little grains and chaff to spread out. It wasn't long before the kernels escaped the piles and got into and onto everything. There was grain in the food, on the chairs, on and under the couch, and on every flat surface. Grain was in every crevice, in their clothes, in their bedding, and in the hair of their dogs and cats. It was everywhere! Each time they turned or stirred the grain, the little hulls and chaff would lodge in their throats and make them choke. The whole household gave up worrying about it, and since it couldn't be helped, decided to just live with it. After all their efforts, their hopes were still destroyed. Kearnel sold only a few tons of the wheat that year. Even that was returned, because it was moldy and unusable.

That was the year endurance was tested beyond anything any of them had ever known. No one was sure exactly how they could go on, but go on they did. There was no money, no crop, and no hope for any of the things they had been dreaming about. As autonomous as the little pioneer family was, they were devastated. Certain items couldn't be grown at home, and they didn't know how they would acquire what they needed. There was a need for sugar, salt, coffee, material for clothing, thread to use for sewing, shoes to keep feet warm in the winter, tools, machinery, seed for next year's crops, and a few other sundries necessary to maintain a home and keep a family.

Sena called upon her great faith. She was a

staunch believer. Nourished by her daily Bible reading, nothing could sway her from the Christian roots that were firmly planted within her. She thanked the Lord for the sunshine, and then thanked Him for the rain with the same devotion. *"The Lord giveth and the Lord taketh away. Blessed be the name of the Lord!"* she said to all, even in the face of gloom, with a bleak and dreary winter looming ahead of them. She praised God in a loud and cheerful voice. To bring hope, Sena read a couple of Bible verses to the family that God had given her in her daily reading.

> *Pray without ceasing. In every-thing give thanks: for this is the will of God in Christ Jesus concerning you. (1 Thessalonians 5:17,18)*

> *Consider the lilies of the field, how they grow; they toil not, neither do they spin: And yet I say unto you, that even Solomon in all his glory was not arrayed like one of these. (Matthew 6:28,29)*

She assured her family, "Man's extremity is God's opportunity. We've just had a little lesson in growing closer to the Lord. Just wait until spring; there will be trilliums, wild roses and daisies everywhere. Money couldn't buy any one of these."

Kearnel's philosophy was an anathema to present

day psychiatry. He approached the trouble at hand by saying, "Okay, we're in trouble and it's going to be a tough year. Now is our chance to see what kind of material we're made of. No one is going to complain. I forbid it! I'll disown you if you do. We're healthy and we're smart. We have this beautiful farm. We've got the pioneer spirit, and if we hang in there together, someday we will be rich."

T.M. Jones, owner of the General Merchandise Store, was one of God's own angels to this pioneer family. He set up an account for K.O. so the household and farm could get what was needed on credit. That was a bitter pill to swallow for an independent man like Kearnel, but it had to be done. One bad year can sink a farmer, or at the very least, put him behind for a long time. The Lund family was no different. After the wheat crop disaster, they had to struggle for years to financially get back to ground level. K.O. always knew he would make good on his debt to T. M., and the latter never doubted for one second that he wouldn't. Kearnel finally began to see the profit margin growing and the debt shrinking once again. He raised some cattle one year and took them to market, receiving a better than fair price for them. After the sale, the first place he headed was to see T.M. to make good on the balance of his debt. He told the store proprietor that he was there to square up the bill.

Mr. Jones looked a bit puzzled and said, ''Why, K.O., you know your bill has been forgiven and written off for years."

But Kearnel insisted, "You trusted me when I

needed help, and an honest man will always pay his bills. Figure up what I owe you, including interest. I'm here to pay."

Remember what Sena had promised her family that hard winter when their dreams of material things were dashed to bits through an act of nature? She had promised flowers in the spring, an act symbolizing God's renewed grace to us. Every time we fall, He will remain true to His promises to us. That will never change. It's just like the disastrous wheat crop harvest that could not stop the flowers from returning. And return they did, in profusion, all the way up Big Creek Trail. Morning glories grew all around the back porch and shaded it from the sun, and the birds sang the sweetest melodies. Sena, standing there surveying God's great glory, began to sing, "Praise God from whom all blessings flow, praise Him all creatures here below." As her beautiful voice rang true and clear, floating skyward, she lifted her family up to God in worship. The soft voices of the little ones joined hers, and soon Kearnel's baritone voice could be heard, blending harmoniously, as their prayers of thanks and praise drifted across the wheat fields and down the valley.

Kearnel and Sena planted many more crops as their lives unfolded throughout the years. The seasons came and went; there was more rain, more wind, more snow, too much sun, and too little sun, all the elements that nature could throw at them. But the most important seeds they ever planted were the ones sown in the hearts of their children, the seeds that rooted deep in the love of God. Because of that, Kearnel and Sena's story continues to be told today throughout the many generations of the Lund families. There have been four generations that have lived and prospered in this rich valley. Their descendants talk of how God provided for the founding family, Kearnel and Sena Lund, and how He is still providing for them today. The forces of nature that were completely out of their control didn't stop Kearnel and Sena from prospering and realizing God's great wealth. They learned that as we come up against unknown and unplanned events that strike against us, it doesn't alter what God can do for those who believe in Him. It is God who ultimately supplies all of our needs. We can be weary from the storms of life that might come across our paths, but always know that God is in control, *yesterday, today and forever.*

Afterword

The "Good Life." It ranks high on our value scale with one minor problem—no one really knows what it is or how to get it.

Some believe it is things: "The one who dies with the most stuff wins." Others think that power, whether political, social, corporate, emotional, or whatever, holds the key. There are those who cast their vote for sex–the more the merrier! Many, of course, hold to the idea that education, knowledge, fame, talent, money, or a combination thereof, will lead to the proverbial pot of gold.

The stories you've just read attest to the fact that the "Good Life" has nothing to do with empty externals. We believe you're in for a great discovery. The door to that discovery is a person. His name is Jesus. He is the source of the authentic "Good Life." Jesus stated emphatically, *"I am the way, the truth, and the life...."* (John 14:6) He wants to be all of that and more to you.

Jesus also said, *"I came so they can have real life, more and better than they ever dreamed of."* (John 10:10; The Message) It's not about mere religion. It's about a brand-new life through a relationship with God by trusting in Jesus Christ. He cares about you and loves you more than you can even begin to imagine.

Scripture reassures us, *"No one who trusts God like this–heart and soul–will ever regret it. It's exactly the same no matter what a person's religious background may be: the same God for all of us, acting the same in-*

credibly generous way to everyone who calls out for help. Everyone who calls, 'Help, God!' will get help." (Romans 10; The Message)

It's really that simple. God is already that close to you. You can begin a new life today by asking Him to show you His truth. Simply make the following prayer the expression of your heart:

> *Jesus,*
> *I am tired of my life as it is. If You can change the people in this book, You can change me. I need You to give me Your "Good Life." I want to know more about You and this life You give. I want to trust You with my heart and soul—now and forever. Please come into my life and take over.*
> *Thank You for making the sacrifice necessary for my sins. Thank You for making it all possible.*
>
> *Amen*

As soon as possible, talk to one of the pastors of the local churches listed on page 5 of this book, and tell them you have chosen a new path, something the Bible calls *"the path of life."* They will assist you in getting started on your new journey, a journey with a future brighter than the most brilliant diamond.

For more information on reaching your city with
stories from your church, please contact
Good Catch Publishing at…
www.goodcatchpublishing.com

GOOD CATCH
PUBLISHING